World War I

Great Battles for Boys

Joe Giorello

with
Sibella Giorello

World War I
Great Battles for Boys

Table of Contents

FOREWORD

Images of WWI

IMAGINE YOU'RE WALKING down the street with your friend. Some other boys are coming toward you.

"I don't like your friend," one of them says.

Then he throws a punch.

You try to stop the fight, but the other boys are all jumping into the clash, throwing their own punches. Before you know it, that first punch has triggered an all-out brawl.

That's sort of how World War I started.

In June 1914, Archduke Franz Ferdinand was shot and killed by an assassin.

Archduke Franz Ferdinand

The archduke was the leader of a country called Austria-Hungary. His assassin—the man who shot him in the neck—was from Serbia, one of the European countries controlled by Austria-Hungary. Along with those other countries, Serbia wanted its independence. But Austria-Hungary refused to allow it. When the assassin killed the archduke, it was sort of like that guy throwing the first punch at your friend.

The fight was on.

This 1914 poster shows an Austrian fist crushing a Serbian holding a bomb and knife. The Austrian phrase translates, "Serbia must die!"

Austria-Hungary declared war on Serbia and formed an alliance with Germany, a country that was already preparing for war. When two or more countries form an alliance, that means they agree to stick up for each other.

Soon after forming this alliance, Austria-Hungary attacked Serbia. The aggression drew the rest of Europe into the fight, forcing these other countries to protect themselves and their alliances. The United Kingdom (England), France, and Russia formed one alliance. That group was called the Allies.

The other side—Germany and Austria-Hungary—was called the Central Powers. Later the Ottoman Empire and the country of Bulgaria would join the Central Powers, too.

As you read about these WWI battles, keep these two groups in mind—Allies and Central Powers—because you'll be hearing a lot about them, and what happened to them during the war.

Look at the map below. It shows Europe in 1914, just after WWI broke out. Find Germany and Austria-Hungary. Then find Russia, France, and the United Kingdom which includes

England (Great Britain), Ireland, Scotland, and Wales. Also notice that some countries, such as Spain, Switzerland, Denmark, Norway, and Sweden stayed "neutral," meaning they didn't take sides in WWI. (You'll also find a list of the Allies and Central Powers at the back of this book).

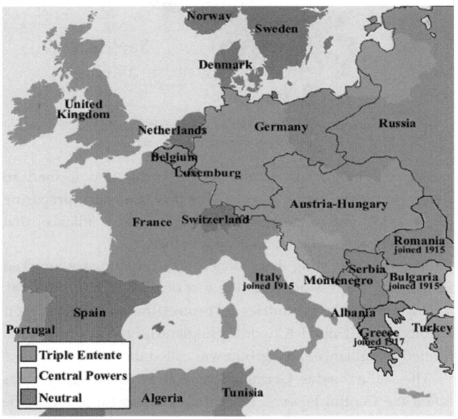

National Archives of Britain

Germany already had a plan for this war. It had long wanted to expand its borders and grow richer by taking control of other countries. To win this new war, Germany planned to strike fast to the west, conquering France within two weeks. Then it would wheel its forces around and fight Russia to the east.

But in this plan to conquer France, the German army first needed to invade Belgium. Look at the map again. Belgium sits between the northern edge of Germany and France.

Belgium had declared itself neutral—it didn't want any part of this war.

But on August 3, 1914, Germany invaded Belgium.

The invasion was brutal.

German troops burned down homes, terrorized people, and killed civilians—ordinary folks who were not soldiers. These actions were all part of Germany's plan to scare the Belgian people into submitting to German authority.

German army invading Belgium, 1914

Great Britain (part of the United Kingdom) threw its support behind Belgium and France. It declared war on Germany.

See how quickly this fight escalated?

Many of the WWI battles you're going to read about take place around Belgium and France, an area that would later be known as the Western Front.

The Eastern Front formed along the border between Russia and Germany.

When the war broke out, most people thought the fighting wouldn't last long. In August 1914, as German troops burst into Belgium, Kaiser Wilhelm II—Germany's leader—assured his soldiers, "You will be home before the leaves have fallen from the trees."

The Kaiser was wrong.

WWI lasted four long years, killing about 10 million military personnel and seven million civilians. The war's casualties—meaning, the wounded—were even higher, about 37 million people total. These almost unbelievable numbers reveal why WWI would later be called "The Great War"—not because it was so good, but because its losses were so big.

How did all of it happen?

Let's find out.

Battle of Mons

August 23, 1914

British cavalry soldiers attack German troops at Mons.

WHEN GERMAN MILITARY forces invaded Belgium in August 1914, their plan was to storm across that country into its neighbor, France. From there, the German army would seize control of France's most famous city, Paris, giving it vast control over important parts of Europe.

To combat the German plan, the French army and the British Expeditionary Force (BEF) set up defensive positions along the border between France and Belgium. This area would later become an important part of the Western Front.

Look at that map again. Notice how small Belgium is compared to its neighbors Germany and France.

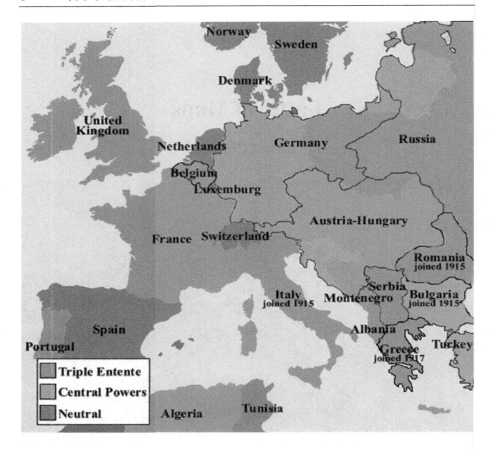

The Allies wanted to halt the German invasion at the Belgian border, then drive the enemy all the way back into Germany. If this Allied plan had succeeded, World War I might've ended right there.

Instead, during this first month of the war, both sides witnessed the new and devastating power of modern weapons and warfare. Machine guns. Artillery. Airplanes. Chemical gas. Tanks. These inventions were going to make WWI different—and much more deadly—than any war that came before it.

Although there were skirmishes during this first month, WWI's first major clash was the Battle of Mons.

Mons is a Belgian mining town with a nearby canal. This

waterway would be really helpful for moving soldiers and supplies around the area. That's why both sides wanted control over Mons and its waterway.

On August 21, the British Expeditionary Force (BEF) was setting up positions to protect the French army that was already in the area. The BEF also sent out a cavalry unit—soldiers on horseback—who rode ahead of the Allied position to scout for Germans. The cavalry hoped to set up an ambush and flush any enemies from the area. An ambush is a trap, usually involving soldiers hiding behind camouflage, such as trees, bushes, and buildings, before pouncing on the enemy in a surprise attack.

WWI French cavalry armed with lances

However, a German cavalry unit spotted the British cavalry, ruining any chance of an ambush.

The German soldiers galloped away.

But the British soldiers chased them, swords flashing, lances raised, rifles firing.

"The enemy fire was hellish," recalled British Second Lieutenant Roger Chance. "We galloped along a black coal dust that blew around us in dark clouds. We could not see a yard in front of us. My horse Spitfire, jumped over two dead horses just in time. Bullets and shells seemed to be everywhere. When we reached a building to shelter, only ten members of my troops were left. I had been struck by a bullet on my metal collar badge but had not noticed it."

The BEF cavalry eventually won this skirmish and took German prisoners.

Now the battle turned to the city of Mons.

WWI British soldiers inside a machine gun trench

The British took up positions along the Mons canal. The waterway's twenty-five miles and twelve bridges made it a huge territory to protect. In order to defend the bridges, the British placed artillery—cannons, machine guns, and artillery shells (explosives)—along the canal. But even with all that firepower, the soldiers struggled to find any good "field of fire" positions—places where they could clearly see the enemy.

Meanwhile, the German army assumed the British were still in France, trying to mobilize—gather together—a military force. Instead, British forces had positioned themselves at Mons—right where the Germans were planning to attack!

Unfortunately, things still didn't look good for the British. At Mons, the Allies had only about 80,000 soldiers and military personnel, plus some 300 artillery weapons. The Germans had around 250,000 soldiers—giving them an advantage of more than three to one. The Germans also had twice as many artillery weapons.

The British did have one big advantage, the Royal Flying Corps. British planes flew over Mons and the pilots spotted the massive German force marching toward the town. One British pilot also tried to warn the French commanders whose forces were nearby, but the pilot's warning was ignored.

WWI French soldiers awaiting German attack.
Notice their distinctive caps, called Kepis.

The British dug in, using the canal and the explosives on the bridges as their defensive line. If the Germans started to break through the line, the British would blow up the bridges, holding back the Germans from crossing the canal.

On August 23, Sunday morning church bells rang out in Mons.

But the calm didn't last.

At 7:30 a.m., German artillery shells smashed into Mons, blowing up both soldiers and civilians. Behind this bombardment, the German army pounded forward with its massive infantry and cavalry force. The British struggled to defend the long defensive line, and soon they were also getting flanked—attacked from the sides.

But the canal's water was too deep for German infantry or cavalry to cross. The only way into Mons was over those bridges.

The 4th Royal Fusiliers were holding the Nimy Bridge, under the command of Captain L.F. Ashburner with nearby support from two machine guns commanded by Lieutenant Maurice Dease.

The German soldiers marched toward the Nimy Bridge in close order formation—the soldiers gathered together to form a concentrated tactical force. But this formation also created good targets for the British soldiers who now aimed their Enfield rifles at them. The British soldiers were so well-trained with these weapons that they could fire thirty rounds in one minute. That meant, every two seconds, the soldiers cocked the gun, fired, locked and loaded. Over and over again.

Nearby, the British machine guns were firing 600 rounds a minute.

The German fire was just as constant, and during the fight, Lieutenant Dease was wounded three times. He refused to leave his crew. Each time one of the machine guns stopped working, Lieutenant Dease would leave his position and go fix the gun, exposing himself to enemy fire. Dease's third wound, however, proved fatal. After the war, he was posthumously—after death—granted Britain's highest military award, the Victoria Cross.

Belgian soldier firing machine gun against the German advance, 1914

The British firestorm of bullets forced the Germans to retreat. However, the Germans learned their lesson. On their next attack they advanced in looser formations, creating less of a target, and they added artillery and planes. German pilots flew up on surveillance—scouting—and marked the enemy's positions on a map. This information gave the ground soldiers even better firing accuracy.

Up and down the Mons canal, all the bridges were under heavy attack. By 3:00 p.m., the Germans strengthened their push with reinforcements. Soon the Germans began to flank, forcing the British to defend both front and sides, stretching their defensive forces even further.

And then, bad news arrived.

The nearby French 5th Army was withdrawing from this battle. The British were ordered to withdraw, too.

But with any withdrawal, an army needs someone to stay behind and cover their retreat. Private Sidney Godley volunteered. Manning a machine gun, Godley kept up a steady fire against the Germans, even after being wounded. His cover gave the 4th Royal Fusiliers time to retreat from the Nimy Bridge.

In order to slow down the pursuing Germans even more, the British blew up five bridges. Over the next two days, British forces held back the German advance, costing the invaders both time and lives. The British also retreated in "good order." That's when a military force doesn't just run away but fights while backing up to leave the battlefield.

The Battle of Mons was over.

The Germans had won.

But this first major battle of WWI proved the British Expeditionary Forces were prepared to fight. The Battle of Mons is also considered one of military history's best retreats. Despite losing the battle, outnumbered and overwhelmed British forces still managed to hold back a German onslaught. The British also protected nearby French forces, ensuring that the Allies lived to fight another day.

But the German plan hadn't changed. They were still fixed on marching into France and conquering Paris.

WHO FOUGHT?

Private Sidney Godley

Sidney Godley joined the Royal Fusiliers in 1909. He was twenty years old and known for being a good soccer player and cross-country runner.

His battalion was among the first sent to France and Belgium, arriving at Mons on August 22. When the battalion reached the Nimy Bridge outside town, French forces were struggling against the German advance. The next morning, the British Expeditionary Force dug into its defensive positions. Godley's job was to help supply ammunition to the position's machine gun.

The Germans nearly destroyed the entire battalion, and when Lieutenant Dease was killed, the troops were ordered to withdraw. Godley volunteered to take over the machine gun to protect the retreat. He volunteered knowing that his choice had two possible outcomes: either he would die in this fight, or the Germans would take him prisoner.

Single-handedly, Godley held the bridge for two hours, giving his fellow soldiers in the Royal Fusiliers time to fully withdraw. When Godley finally ran out of ammunition, he broke up the machine gun and threw the pieces into the canal, keeping the weapon out of German hands.

Despite a bullet lodged in his head, Godley managed to crawl back from the bridge to a main road. Two Belgian civilians dressed his wounds, but the Germans captured the post and took Godley prisoner. Although questioned heavily, Godley gave the Germans only his name, rank, and number.

Godley was sent to Berlin, Germany and underwent surgery for the bullet in his head and for injuries to his back—injuries that required 150 stitches. He was then transferred to a Prisoner of War camp.

In 1919, after the war ended, Godley was released and returned to England.

Private Sidney Godley was the first WWI *private* to receive Britain's highest military honor, the Victoria Cross.

BOOKS

Journal of William Brazear: An Eye Witness Account of the Battle of Mons by William Brazear

Mons 1914: The BEF's Tactical Triumph by David Lomas

INTERNET

Documentary on the Battle of Mons:
youtube.com/watch?v=oEFoZsuLRoE

The Battle of Tannenberg

August 23–30, 1914

Russian troops awaiting German attack

AT THE BEGINNING of WWI, the Germans didn't have much respect for the British army—or the Russian army.

Russia is a huge country with many resources, but it wasn't well prepared for modern warfare. When WWI broke out, Russia didn't even have enough rifles for all its soldiers. Or artillery.

Russia also didn't have enough communication wiring. Unlike today's wireless networks that use satellites and other technology, in the early 1900s people used telegraphs and radios to speak directly to each other over long distances. Without wiring to connect these devices, people had to resort to slower communication methods, such as sending letters by mail, or messengers with notes, or even carrier pigeons.

These negative factors were part of the reason Germany felt confident about defeating the Russians.

Making matters even worse, the Germans had also broken the Russians' secret code. Any time a top-secret message was sent among the Russians, the Germans could read it and find out exactly what the Russians planned to do. This code-breaking gave the Germans a huge advantage in battle.

Still, the Russians managed to move their 1st and 2nd Armies into the WWI battlefield sooner than expected. By mid-August, Russian soldiers were at the Eastern Front. Now they would face Germany's fierce 8th Army.

Look at the map. The Eastern Front ran along the border between Russia and Germany, concentrated mostly in that deep U-shaped curve that bulged into Germany.

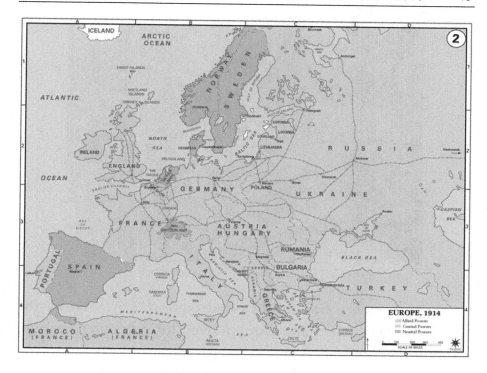

The Russians planned to capture the German 8th Army in a pincer movement. That's when an army sends out two flanking forces, left and right. Those forces create something shaped like an open claw. As the claw closes, the enemy is trapped inside.

Only problem was ... the Russian forces had yet *another* problem.

The Russian territory was so vast and the Russian armies so large that the military struggled to keep its soldiers supplied with food and ammunition as it moved them from place to place. There were also thousands of horses that needed grain and hay. Without supplies, the troops would be forced to forage—live off the land—something that's very hard to do with hundreds of thousands of men and animals.

WWI Russian soldiers in a forest trench

After a series of skirmishes that were mostly won by the Russians, the Russian 2nd Army moved forward, pressing into the outnumbered German lines.

But instead of joining the 2nd Army, the Russian 1st Army stopped to rest.

This decision caused a serious problem (or you might say, *another* serious problem) for the Russians.

It would also lead to one of the worst disasters of WWI, the Battle of Tannenberg.

As the two Russian armies separated, the Germans decided to move their main force against the 2nd Army.

At first the German plan was just to hold back the 2nd Army. But on August 23, several new German generals took control and they decided to attack with the 2nd Army full throttle. General Erich Ludendorff later explained his reasoning.

"Our decision to give battle arose out of the slowness of the Russian leadership and was conditioned by the necessity of winning in spite of inferior numbers…"

The Russian 2nd Army was marching about ten miles a day. Supply lines struggled to keep up that pace. The Russian soldiers were tired and growing hungrier by the day.

The Germans, on the other hand, sent some of their forces ahead by railroad, which was a lot less tiring than marching over ground. Another advantage for the Germans was the land where the Russians were headed. Full of lakes, swamps, and forests, it provided excellent coverage for hiding and ambushing. And finally, the Germans knew this country really well because they had used it as their training grounds.

Russian troops marching to the Eastern Front

The German army closed in around the Russian 2nd Army, and the Russians had no idea it was happening. Basically they were marching into a very clever trap—a trap that was almost identical to the pincer movement the Russians had planned to use on the Germans!

Look at the map below. It shows how on August 20, the Russian 1st Army had stopped to rest in the north (upper right corner). The Russian 2nd Army was far to the south (lower right). Between them, the German forces led by Ludendorff was coming by railroad.

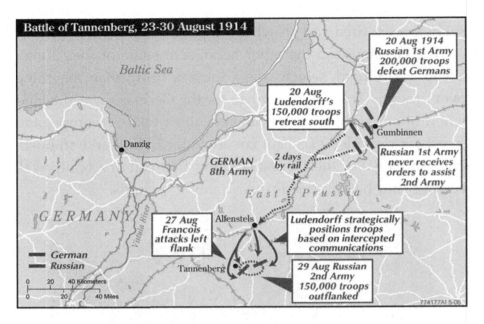

Battle of Tannenberg, 23-30 August 1914

Baltic Sea

20 Aug 1914
Russian 1st Army
200,000 troops
defeat Germans

20 Aug
Ludendorff's
150,000 troops
retreat south

Gumbinnen

Danzig

GERMAN
8th Army

2 days
by rail

Russian 1st Army
never receives
orders to assist
2nd Army

East   Prussia

GERMANY

27 Aug
Francois
attacks left
flank

Alfenstels

Ludendorff strategically
positions troops
based on intercepted
communications

German
Russian

Tannenberg

29 Aug Russian
2nd Army
150,000 troops
outflanked

0   20   40 Kilometers
0   20   40 Miles

774177AI 5-06

In late August, German gunfire burst through the forest. Black smoke choked men's lungs. German machine guns spewed a seemingly endless supply of bullets. Russian flanks started retreating. Although the Russian center managed to make some advances, the Germans had dug into trenches and were using the terrain's excellent cover to keep the Russians from gaining much ground. Finally, after realizing the oncoming slaughter, the Russians broke and ran—retreating so quickly that soldiers dropped their rifles.

Germans in machine gun trench

As German artillery shells burst around them, Russian soldiers used their hands to dig into the hardened dirt, hoping to find cover. Wounded men screamed as they fell. Bullets hailed overhead.

The Germans refused to let up.

Climbing from their trenches, they went in for close-combat fighting. While some Russian soldiers raised their hands in surrender, others raced into the woods, hoping to escape.

German General Ludendorff later described the scene:

"As we at the head of the column came out of the dreadful wood, a shower of infantry fire suddenly hailed down on us. Lieutenant Colonel Schultz stopped a bullet in the temple and fell like a board, but he soon came to, swore frightfully and asked for a cigarette. Meanwhile, we had brought up artillery from the wood, and the Russian rabble, leaving behind a number of rifles and packs, beat a hasty retreat, back into the darkness from which it had emerged."

The Germans destroyed Russia's 2nd Army.

The Battle of Tannenberg was over, and the Germans had won—decisively.

The Tannenberg defeat was so humiliating for the Russians that the commander of the 2nd Army, General Alexander Samsonov, walked into the nearby woods, pulled out his revolver, and killed himself.

Russian prisoners at Tannenberg

In the battle, Germany lost 20,000 men, dead or wounded. Russia lost about 140,000 men, dead or wounded. Another 150,000 Russian soldiers were captured.

Hearing the news, Russia's 1st Army—so far away it couldn't help its fellow soldiers—was ordered to retreat.

WHO FOUGHT?

Russian Cossacks serving as soldiers in WWI

The Cossacks were among the most legendary fighters of WWI.

Tall and strong, the Cossacks were excellent horseback riders. They lived on the far borders of Eastern Europe and fought on behalf of Russia during the war. Usually the Cossacks were tasked with scouting ahead of the advancing army or with covering an army's rear.

The Cossacks were armed with M1891 carbines (a type of rifle). They carried thirty rounds of ammunition in two oilskin bandoliers—a belt with pockets, often slung over a shoulder. But even with that weaponry, the Cossacks preferred hand-to-hand fighting—which was one reason their enemies found the Cossacks so terrifying. In close-combat battles, the Cossacks used a saber sword called a shashka.

French military general Napoleon Bonaparte once described the Cossacks as "among the best light troops among all that exist. If I had them in my army, I would go through all the world with them."

But ironically at Tannenberg, the Cossacks were among the factors that put a strain on the Russian forces. The Russians sent a large number of Cossacks to fight the Germans on the Eastern Front.

But horses require even more food than people.

BOOKS

World War I: A definitive visual history by DK

DK Eyewitness Books: World War I by Simon Adams

INTERNET

Quick video with footage of WWI cavalry, including the Cossacks on the Eastern Front. Also information about wolf packs attacking wounded soldiers! youtube.com/watch?v=XovnkqJaqL8

MOVIES

History's Great Military Blunders and the Lessons They Teach (2015 video)

Aviation and Aircraft

Source gallica.bnf.fr / Bibliothèque nationale de France
Early attempt at mounting a machine gun on a WWI-era plane

BATTLES HAD ALWAYS been fought on land or water—until World War I.

In 1904, ten years before the war broke out, the Wright Brothers launched their first short flight over Kitty Hawk, North Carolina. Soon after, countries around the world started developing airplanes and men began learning how to fly them.

However, in the early days of WWI, military commanders considered airplanes useful only for scouting and surveillance—gathering information on the enemy. Pilots would fly over enemy territory, sometimes with a copilot, and observe an enemy's position along with whatever weaponry and other supplies they might have. This crucial information was sent back to the ground forces. Sometimes that information swung the battle. In the Battle of Tannenberg, for instance, Russian General Alexander Samsonov ignored the information gathered by his pilots. His entire army was either killed or captured (and as you now know, Samsonov committed suicide). Meanwhile, German commanders paid close attention to the information uncovered by their pilots and decisively won the battle. One German commander later said, "without airmen there would have been no Tannenberg."

Downed WWI fighter plane

But for all the flying these pilots were doing, you might be surprised to hear how little training they received. Today, military pilots train for thousands of hours before ever seeing combat. But in WWI, some men had barely flown their plane before being sent into combat missions. With the pilots so inexperienced, early fighter planes were designed for stability instead of speed or quick movements.

Giant balloons provided another way to scout from the air. Cigar-shaped vessels called zeppelins, the balloons were filled with various gases, including helium. The gasses were lighter than air, causing the balloons to rise. But when WWI broke out, the United States stopped exporting helium gas to Germany. The Germans then switched to using hydrogen gas. However, hydrogen is highly flammable. Despite the risk of fire or explosion, the Germans continued to use hydrogen-filled zeppelins for bombing missions. The first German zeppelin to bomb London was on May 31, 1915.

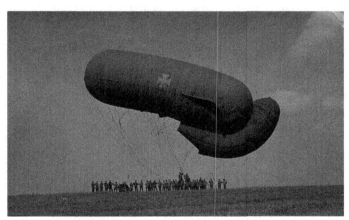
WWI German observation balloon

But by 1915, airplanes became a regular part of the WWI battlefield. Pilots learned new skills, allowing the planes to be designed for speed and agility. Guns were soon added to the planes and "dog fights" broke out in the skies over the battle-field as enemy pilots tried to shoot each other down.

At first, any pilot who shot down five or more enemy air-craft was called an "ace." As the planes and weaponry improved, however, "ace" was used only for downing ten or more aircraft. The British had the most aces—785 pilots. The

Germans came in second with 363 aces. And France had 158.

To fly combat, WWI pilots needed real courage. Their job was so dangerous that the life expectancy of a fighter pilot was—are you ready?—two weeks.

But as the war continued, these pilots and their airplanes became an essential part of modern combat. With aircraft, no enemy territory was off-limits. Pilots could fly to the most remote islands, above high mountains, and, as technology improved, stay airborne for hours at a time, scouting an army's entire supply line. Armed with guns and bombs, pilots could also wipe out armies on the ground.

The British developed one of WWI's best fighter planes, the Sopwith F-1. Nicknamed the Camel, the Sopwith was considered fast for its time. It also came equipped with a stable gun platform that didn't wobble when a pilot flew up on an enemy. The platform gave the pilot's machine gun better accuracy.

German airplane downed near the Western Front

The Germans developed their own innovative plane, the Fokker D.VII. This plane combined speed, agility, and firepower, all built around a steel tube frame. While other WWI planes had wooden frames, the Fokker's metal frame helped German pilots survive more mid-air dog fights and crash landings.

Most machine guns were mounted in front of the plane for forward firing. But consider this: these planes were powered by a propeller on their nose. Sometimes pilots would accidentally shoot off their own propeller blades trying to take down an enemy! To solve this problem, the Germans developed a synchronized mechanism. It timed the machine-gun fire so that the bullets would fire only between the spinning propeller blades.

Eventually WWI fighter pilots needed to master bombing skills, too. At the beginning of WWI, French pilots were dropping six-inch-long steel darts capable of piercing a man's skull. But these darts weren't very accurate. Meanwhile, the Germans invented a bomb with a propeller that activated its own fuse. But it also wasn't very accurate. By 1916, however, the Germans developed a torpedo-shaped bomb that was a lot better at hitting its target.

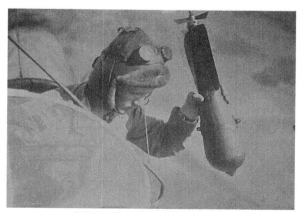

WWI pilot aiming his bomb.

Most of WWI's medium bombs weighed about 100 pounds. But the British dropped WWI's largest aerial bomb. That thing weighed about 1,600 pounds—about the same weight as four modern automobiles!

Cockpits in these early fighter planes were very small. In fact, the space was so tight that Allied pilots didn't wear parachutes. Allied commanders also reasoned that if their pilots had parachutes, they might bail out of the plane at the first sign of danger.

These early planes also didn't come with windshields, or any other protection from the whipping air, freezing cold, and deadly enemy bullets. Pilots began wearing silk scarves to keep their neck muscles warm, allowing them to keep turning their heads as they scouted enemy territory and watched for enemy planes.

By late 1915, airplanes had become so essential to the war that Allied manufacturing went into heavy production. For example, at the start of WWI, France had fewer than 140 aircraft. By the end of the war, just four years later, the French air fleet had more than 4,000 planes—and that number didn't include all the planes that were lost or shot down during the war.

WHO FOUGHT?

The Red Baron

Have you ever read the comic strip *Peanuts*?

If you have, you've probably seen Snoopy sitting on top of his dog house pretending to be a WWI flying ace as he yells, "Curse you, Red Baron!"

The Red Baron was a WWI flying ace.

Born Manfred von Richthofen, he enrolled in military school at age 11. Later he became a cavalry officer. He also served in WWI trenches for a brief time. In 1915 he was transferred to the German Air Force.

Richthofen quickly mastered the necessary flying skills. After his twenty-four hours of flight training, he took his first solo flight. Soon he had scored six "kills" against Allied aircraft.

Richthofen was now an ace.

Although level-headed and precise in battle, Richthofen painted his Fokker Dr.1 Dridecker a bright red color. As his kill rate in the air grew famous, the British nicknamed him the "Red Baron."

In 1917, Richthofen was commanding Germany's top fighter pilots, a group known as the Flying Circus. These pilots were sent all over the Western Front where they caused catastrophic damage to the Allies. Richthofen himself was still flying into battles, scoring 80 confirmed kills, but was finally shot down while flying deep into British territory.

There's some controversy about who actually shot down the Red Baron. Some people claim it was a Canadian pilot. Others insist the deadly shot came from some Australian gunners in trenches on the ground.

Either way, Manfred von Richthofen crash-landed into a field and died from his injuries. He was 25 years old.

A British pilot flew over Germany and dropped a note informing the Germans that their legendary fighter pilot was dead.

BOOKS

The Red Baron: The Graphic History of Richthofen's Flying Circus and the Air War in WWI (Zenith Graphic Histories) by Wayne Vansant

The Red Baron: The Life and Legacy of Manfred von Richthofen by Charles River Editors

The Red Baron and Eddie Rickenbacker: The Lives and Legacies of World War I's Most Famous Aces by Charles River Editors

INTERNET

Vintage film footage showing WWI aircraft and some dogfights: youtube.com/watch?v=dwrIf_5gEEM

Here's a cool WWI aviation timeline from the US WWI Centennial Commission: www.worldwar1centennial.org/1181-timeline-of-wwi-aviation-history-demo.html

MOVIES

Dog Fight: The Mystery of the Red Baron (2017)

The Battle of Gallipoli

February 19, 1915 – January 9, 1916

Australian troops charging from a trench at Gallipoli

WINSTON CHURCHILL WAS among the greatest leaders of England—of all time.

During World War II, Churchill's brilliant plans would help stop Hitler from taking over the world.

But Churchill wasn't perfect.

In fact, he made some really big mistakes—especially during World War I.

During the war, Churchill was head of Britain's navy—called the "First Lord of the Admiralty." Although Churchill came up with a smart plan to win WWI, it actually turned into a colossal tactical error that cost thousands of men their lives.

Welcome to the Battle of Gallipoli.

Here's how Churchill's plan started.

In 1915, the Western Front had fallen into a stalemate. Neither the Allies nor the Central Powers was winning the war. Instead, soldiers on both sides stayed hunkered down in muddy, disease-ridden trenches as a commander blew his whistle, sending the soldiers racing from their trenches into an area called "no man's land."

No man's land was located between the opposing trenches. Often strung with barbed wire, it was where hundreds of thousands of soldiers were slaughtered running toward deadly enemy gunfire. Day after day, this trench warfare continued on the Western Front, wiping out the Allied troops—without bringing any victory into sight.

Something more was needed to stop the Central Powers.

Infantry from the British Royal Naval Division race from trenches
at the Battle of Gallipoli, 1915.

Although Russia was part of the Allies, it was experiencing so many problems at home that it was struggling to fight the Germans, too. Russia's leader, Czar Nicholas II, was now pleading with the Allies for help on the Eastern Front.

Churchill's plan was supposed to provide that help. And the plan seemed sound at first, in part because it was based on an earlier successful military mission. On November 1914, Churchill had ordered an Allied naval expedition to sail into a narrow waterway called the Dardenelles. This area was part of Turkey, which was controlled by the Ottoman Empire (more about the Ottomans later). When English battlecruisers in the Dardenelles fired on Turkish forts, one artillery shell hit a magazine—where ammunition is stored. The magazine's explosion destroyed ten guns and killed 86 Ottoman and 40 German soldiers—an Allied success.

Allied fleet, sailing into the Dardenelles.

Now Churchill wanted to return to the Dardenelles.

Look at the map. It shows Europe in 1914. At the bottom right corner, you'll see Turkey.

Notice also the waters around Turkey—such as the Black Sea—some of which spread toward Russia.

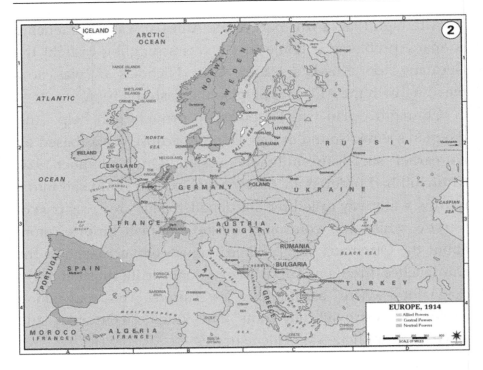

That earlier attack on the Dardanelles was an Allied success. But it also alerted the Ottomans that they needed to fortify their defenses.

Churchill's new plan called for Allied warships to blast those Turkish fortifications again, only this time, the Allies would inflict so much damage that Turkey would be forced to sue for peace—that means, surrender in good terms.

"A good army of 50,000 and sea-power—that is the end of the Turkish menace," Churchill said.

If the Allies could force Turkey to drop out of the Central Powers, the loss would put more pressure on the Germans. Then, with control of the Dardanelles and the nearby seas, the Allies could use the waterways to send supplies to the Russian forces fighting on the Eastern Front.

If the Allies could win the Dardenelles, they would gain a powerful strategic advantage over the Central Powers.

British battleship *HMS Canopus* fires a salvo from her 12-inch guns, bombarding Turkish forts. Dardanelles, March 1915

On paper, Churchill's plan looked clever.

But military battles are fought on the ground, in the air, and on the water.

In the Dardenelles, the Ottoman Turks owned the high ground. They placed their artillery so that it fired down on any enemy naval fleets that entered the channel. The Turks also added explosive mines to the waterway. In a narrow channel like the Dardanelles, these mines were especially deadly because ships didn't have much room to sail around the explosives, or to escape.

Here's a map of the Dardanelles. Those dark hash marks show the minefields. Boxes represent Turkish forts.

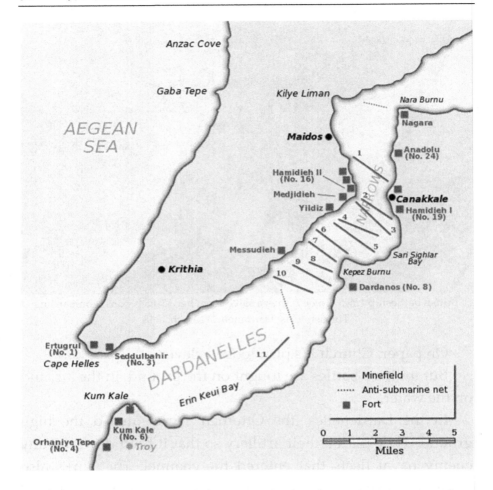

Sure enough, on March 18, 1915, the Ottomans managed to sink three Allied battleships in the channel. The Turks also severely damaged three more ships and inflicted 700 casualties on the British-French fleet.

The Turkish bombardment was so deadly that one Australian officer later described feeling as if, "The key was being turned in the lock of the lid of hell."

The Allied plan now shifted. Instead of attacking from the water, Allied ground troops would be sent in. Their mission? Wipe out the Turks.

The troops were sent to Gallipoli, a peninsula—or narrow strip of land—just west of the Dardanelles.

HMS Irresistible, destroyed by a floating mine explosion in the Dardanelles, March 18, 1915.

These Allied troops landing at Gallipoli were a mix of British, Australian, New Zealand, and French soldiers. The Australian and New Zealand armies were together known as the "Anzac."

The first landing took place on April 25. Allied soldiers were sent to five different beaches. Three of the beaches received only light resistance from the Turks.

The other two landings were a different story.

The Turk's heavy machine-gun fortifications and barbed wire created a death trap. Even worse for the Allied soldiers, the landing vessels were slow and offered little protection for the men inside. In fact, some of these vessels were just old riverboats equipped with little more than a wooden plank that dropped the soldiers so far from their intended landing that men struggled to reach the beach—all the while exposed to a long "killing zone."

The Turks, with strong defensive positions on high ground, blasted machine guns, artillery, and rifles at the Allied soldiers.

These landings were a near-slaughter.

Irish soldier Henry Hanna was among the lucky ones. He reached the beach. But his situation didn't improve from there.

Anzac sniper and spotter in trench at Gallipoli, 1915

"Then came my dash for safety," Hanna wrote. "I got a splinter of a bullet in the side. It just pricked the skin and stuck in my belt. There is a hole in my belt where it stuck. When I got behind the line, the first thing I saw was [fellow soldier] Lex, bandaged all over the head and shoulder, but could see no one else there. There were no stretcher-bearers of any sort so I got permission from Lieutenant Hamilton to help him down the ridge. I then discovered my knee was cut and swollen. Either another splinter of a bullet or cut by the rocks. I could hardly walk. The sights I saw going along that place I shall never forget. Some of our fellows throwing back the bombs which the Turks threw over and which had not exploded. One fellow caught them like catching a cricket ball. Wounded and dead lying everywhere. The sun streaming down and not a drop of water to be had. Neither had we bombs to reply to the Turks and drive them out."

As the Gallipoli landing was turning into a bloodbath, still more problems arose. Not only was there poor organization between the landing troops and the Allied ships in the waterway, but the naval firings also weren't powerful enough to slow down the Turks blasting away at the beachheads. In one section, about 200 Allied soldiers ran for the beach. Twenty-one survived.

Now the Gallipoli mission was looking impossible—even to the Turkish soldiers also getting killed in this battle.

"I don't order you to attack," one Turkish commander told his men. "I order you to die."

The Allied plan was supposed to link up these five different beach landings. Then there would be one powerful force, giving the Allies a six-to-one advantage over the Turks.

But under all this heavy fire, with men struggling to cross the area's steep and rugged terrain, with bad communications and poor naval coverage, the landings never managed to link up. The numerical advantage was lost.

By May, the Turks went on the offensive. About 40,000 Turkish soldiers attacked the Anzac forces, with the goal of driving the Allies back into the sea.

Instead, the outnumbered Anzacs wiped out 10,000 Turkish enemies.

Royal Irish Fusiliers in the trenches at Gallipoli.
Photo courtesy of Australian War Memorial

As the months of deadly fighting dragged on, the Allied forces were soon stuck in the same trench warfare as the Western Front. The machine-gun fire from the Turks never let up. Grenades fell with the regularity of rain. And any man who dared leave his trench was quickly clipped by Turkish snipers. Dead bodies were left where they fell, rotting on the ground. And when rain fell, it came with such force that the trenches flooded, drowning men inside.

Nowhere was safe at Gallipoli.

By August, the Allies had developed another plan.

This time, they would flank the Turks at two locations, Sidi Bahe Ridge and Suvla Bay.

The dual attacks looked promising, at first. Allied soldiers managed to catch the Turks by surprise. But the opportunity for victory was soon lost when poor leadership created delays, allowing the Turks to place reinforcements into position and halt the Allied advance.

Now, with tens of thousands of soldiers dead, a new Allied commander was sent to Gallipoli. His name was Sir Charles Moore, and he soon realized this situation was dire. Even with reinforcements, Moore knew the Allies could not beat the Turks at Gallipoli.

On December 7, 1915, Moore ordered an Allied withdrawal.

The Battle of Gallipoli was over—a profound Allied loss.

During the nine months of fighting, the Allies had suffered about 220,000 casualties. Some 46,000 men were dead.

The Turkish losses were even higher with 250,000 casualties and 65,000 dead.

But of all the troops serving in Gallipoli, perhaps the deepest mark was felt by the Anzacs—the troops from Australia and New Zealand—who suffered some of the most horrendous losses.

To this day, both Australia and New Zealand commemorate the Gallipoli landings. Every year on April 25, both nations celebrate Anzac Day, a tribute to the sacrifices made by their WWI soldiers.

WHO FOUGHT?

Winston Churchill

Long before World War I began, Winston Churchill sensed conflict was coming.

To prepare for German aggression, Churchill, the First Lord of the Admiralty, established the Royal Naval Air Service. He also modernized the British naval fleet and helped invent one of the world's earliest tanks.

But after his failure planning the Battle of Gallipoli, Churchill resigned from the navy. For several years, he worked various government jobs.

But almost as soon as WWI ended, Churchill once again warned about German aggression—what would become World War II.

In 1933, when Adolf Hitler rose to power in Germany, Churchill increased his warnings that Hitler and his Nazi political party meant trouble. However, after all the suffering caused by WWI, most people didn't want to hear any more talk of war.

Churchill's warnings were ignored.

Then in 1939, Hitler invaded Poland—the invasion that launched WWII.

The following year, in 1940, Churchill became prime minister of Great Britain. Throughout WWII, Churchill would prove himself among the strongest and smartest Allied leaders. In fact, without Churchill, many people believe Hitler might have succeeded in conquering England, if not the world.

"I have nothing to offer but blood, toil, tears and sweat," Churchill said in his first speech as prime minister during WWII.

"You ask, what is our aim?" he continued. "I can answer in one word: It is victory, victory at all costs, victory in spite of all terror, victory, however long and hard the road may be; for without victory, there is no survival."

Churchill's nickname became "The British Bulldog."

Interestingly, his failure at Gallipoli wound up playing a role in his later success.

On June 6, 1944, WWII Allied forces landed in Normandy, France. Known as D-Day, this amphibious mission was daring and bold—and bore some similarities to the landings at Gallipoli. Allied soldiers were sent to five different beaches with the goal of linking up later to attack German forces. Despite substantial loss of life, the D-Day invasion was a success and became a turning point in WWII, leading the Allies to victory over Hitler.

Perhaps Churchill had learned some crucial lessons from his mistakes at Gallipoli.

Today, the "British Bulldog" who failed during WWI also ranks among history's greatest wartime leaders.

As Churchill himself once said, "Never, never, never give up."

BOOKS

Gallipoli 1915: Frontal Assault on Turkey by Philip Haythornthwaite

50 Things You Should Know About the First World War by Jim Eldridge

Trenching At Gallipoli by John Gallishaw

The Pals at Suvla Bay (1917) by Henry Hanna

INTERNET

WWI footage from the Battle of Gallipoli:
   youtube.com/watch?v=DxTKqldkyik

MOVIES

Gallipoli (1999)

Gallipoli (2005) documentary

The Ottoman Empire

Bundesarchiv, Bild 146-1970-073-17
Foto: o.Ang. | April 1917

Ottoman cavalry unit, 1917

THE TURKS WHO fought at the Battle of Gallipoli belonged to something called the Ottoman Empire, which joined the Central Powers.

But you might be wondering, what was the Ottoman Empire?

It was founded in 1299 by Osman I, a leader of some Turkish tribes who roamed the land. The term "Ottoman" comes from Osman's name. In the Arabic language, his name is "Uthman."

Ottoman cavalry charge, WWI

The Ottoman Empire was based on the religious teachings of Islam. For more than 600 years, the empire conquered and controlled large parts of the Middle East, Eastern Europe, and North Africa. During these early military campaigns, the Ottoman Turks earned a reputation for being ruthless warriors who gained political power through invasions and battles.

In 1453, the empire's leader, Mehmed II the Conqueror, seized the city of Constantinople. That city was the capital of the Byzantine Empire, which was Christian. But after conquering it, Mehmed renamed the city Istanbul, meaning "the city of Islam." Istanbul became the capital of the Ottoman Empire, marking the end to the 1,000-year reign of the Byzantine Empire.

By the 1500s, another ruler was commanding the Ottomans, Suleiman the Magnificent. Suleiman was a very clever military strategist. He led his large army into many bloody battles and won most of them, expanding the Empire to its largest size, most wealth, and greatest power.

Look at the map. It shows the territory ruled by the Ottoman Empire in the year 1683. (Notice, however, the capital city of Istanbul is still marked "Constantinople.")

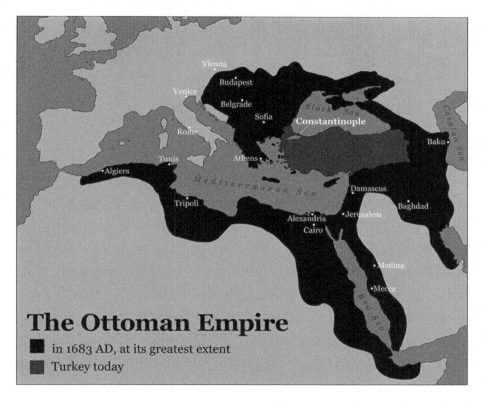

The Ottoman Empire

■ in 1683 AD, at its greatest extent
■ Turkey today

But no Ottoman ruler was ever really safe. The Ottomans practiced something called fratricide ("fra-treh-side"). Whenever a new sultan took over the throne, his brothers were immediately put in prison—just in case they tried to kill him and take his crown. Then the sultan's son, the sultan's brothers and their sons were killed—again, to protect the sultan's power.

Although not every Sultan followed the rule of fratricide, all sultans slept in a different place every night as a safety measure to avoid assassination.

German Emperor Wilhelm II (second from left)
and Ottoman officers at Gallipoli, 1917

Another Ottoman practice was the taking of "Janissaries."

In the 14th century, Christians living inside the Ottoman Empire were required to pay a special tax to the sultan. The Christians were ordered to give some of their strongest male children to the government. These children became slaves, were forced to convert to Islam, and were heavily trained in warfare. They became known as "Janissaries" and became some of the Empire's most passionate fighters on the battle-field—especially when fighting against Christians.

But around this same time, the Ottoman Empire was losing some power.

The parts of Europe that remained Christian were experiencing the Renaissance—a cultural explosion of new ideas, inventions, and improved standards of living. This part of the world was also growing rich through expanded trade routes to faraway countries such as India.

In 1683, the Ottoman Turks marched across Europe to invade the wealthy city of Vienna, Austria (after first trying in 1529). But European Christians inside Vienna defeated the Islamic Turks, and the loss marked the beginning of the end for the Empire.

By the 1800s, many countries under Ottoman control tried to overthrow the Turks—their people wanted freedom and independence. During a conflict known as the Balkan Wars that took place from 1912–13, the Ottoman Empire lost almost all its European territories.

Right after, World War I began.

The Ottoman Empire joined the Central Powers, but in the end the Allies won the war. As the winners, the Allies divided up the Empire's land among themselves.

In 1923, Turkey declared itself an independent republic.

The Ottoman Empire was finished.

WHO FOUGHT?

Ottoman Archer

Turkish Warriors started their military training in childhood.

Around age four, boys were taught how to ride horses. Next came archery—especially learning the skill of firing a bow and arrow while riding on horseback. This cavalry skill would later prove deadly to enemies.

Turkish warriors usually carried their bow slung over one shoulder, like a backpack.

Archeologists have found other close-combat weapons buried in Turkish graves, such as short swords and spears, knives, daggers, and whips. But Turkish cavalry warriors didn't carry their swords where anyone could see them. Instead, they hid them inside packs hung from their belts.

Turkish warriors also carried light shields. But they wore very little heavy armor, partly because their style of warfare relied on speed and agility, and partly because they lived in hot climates where heavy metal armor could cause their bodies to overheat.

In battle, the Turks relied on their cavalry's quick actions and lightning-fast movements. Soldiers on horseback could maneuver much faster than infantry on the ground. During attacks, the Turks were known to scream as a way to terrify their enemy, and to fake withdrawals to confuse them. The Turks also encircled entire armies, trapping them before destroying them.

BOOKS

Great Battles for Boys: Ancients to Middle Ages by Joe Giorello. Covers several Ottoman battles, including the historic Battle of Vienna.

The Ottoman Empire: A Captivating Guide to the Rise and Fall of the Turkish Empire and its Control Over Much of Southeast Europe, Western Asia, and North Africa by Captivating History

Days of Danger by Fritz Habeck

Vienna 1683: Christian Europe Repels the Ottomans by Simon Millar

INTERNET

Video demonstrations of the Turkish warriors' special bow and arrow: defense-and-freedom.blogspot.com/2011/04/exotic-ancient-weapons-i-majra.html

The Battle of Verdun

February 21 – December 18, 1916

French soldiers crawl through barbed wire at the Battle of Verdun, 1916.

WE'RE GOING TO jump forward in time to the year 1916.

WWI was now two years old. Although devastating battles were being fought, the Western Front remained at a stalemate. Neither the Allies nor the Central Powers were making progress, but hundreds of thousands of soldiers were being wounded or killed. At that time, the standard operating procedure (SOP) for armies was to send soldiers forward in waves. The first three waves of soldiers were sacrificed to heavy fire—most of them killed or wounded—before a fourth wave raced forward hoping to overwhelm the enemy. This four-wave tactic killed thousands upon thousands of men in WWI, while achieving only mixed success at best.

Trench warfare was a miserable—and mostly deadly—way to fight.

French soldiers in a trench at Verdun

Imagine for a moment you're a soldier inside one of these trenches on the Western Front. You've been standing in mud and water for so long that your feet never get dry. Horrible smells hang in the air. The trenches also don't have enough latrines—restrooms—so sicknesses spread quickly in the confined and filthy space. Your ears ring from the constant blasts of machine guns and artillery explosions. If you poke your head up, snipers take shots. Sometimes a bitter odor floats toward you triggering a sense of panic—the enemy's chemical gas attacks will burn and blister skin, lungs, and eyes, leaving men blind for the rest of their life.

But when your commander blows his whistle, you must go

"over the top." Scurrying up the sides of the deep muddy trench, you make it onto flat ground. Your bayonet is fixed on your rifle as you race across no man's land, running through a hail of bullets from the enemy. You try not to trip over the barbed wire or the holes left in the ground from exploding artillery shells, but you stumble over dead bodies that lay in grotesque positions. And if you fall, the man next to you can't stop to help. The commander's orders are firm—keep going, no matter the cost.

On both sides of the war, soldiers were facing the same destructive tactics of trench warfare. And because of it, both sides were losing men in almost unbelievable numbers—thousands upon thousands upon thousands.

The deadly stalemate on the Western Front needed to stop.

The Germans came up with a new plan to do just that, and it led to one of WWI's longest clashes, the Battle of Verdun.

German soldiers at Verdun with one of their "rail long guns."
This weapon could hit a target twenty miles away.

The Germans wanted to force the French into using up all their reserve soldiers. Without enough men to keep fighting in the war, the French would be forced to surrender. And without the French, the Allies might soon crumble, too.

The French Army had about 50 divisions, or around 500,000 men—a division is roughly 10,000 to 15,000 men. The Germans had about 80 divisions—close to a million men—giving them a two-to-one advantage over the French.

But to destroy as many French soldiers as possible, the Germans chose to launch a major offensive on Verdun, France.

Why Verdun?

Located in northeastern France near Belgium, Verdun had two main forts, Fort Douaumont and Fort Vaux. At the beginning of WWI, these forts were heavily defended by the Allies. But in 1916, Verdun was stripped of nearly all armaments. The guns were sent to other battlefronts that were in desperate need of artillery.

Now only about 30,000 men were defending Verdun.

Fort Douaumont, Verdun, France

Another attraction for the Germans was Verdun's isolation. There was only one road leading into and out of town, making it even more challenging for the Allies to send in men and supplies. There was, however, a railway nearby and the

Germans planned to use it, quickly delivering their own men and supplies, including ten divisions of soldiers, 1,400 heavy guns—or cannons—and ravaging destruction of flamethrowers.

The Germans dubbed this battle "Operation Judgement."

French army 370mm mortar

During the first two years of the war, the people living in Verdun and the soldiers guarding the forts felt fairly safe. The city hadn't seen much action.

But on the morning of February 21, 1916, German artillery shells screamed across the sky. The shells smashed into the city, blowing up entire buildings. The bombardment continued for twelve hours. By the end of that first day, the Germans had hurled two *million* shells at Verdun.

An Allied pilot who flew over Verdun after this bombardment later recalled, "Every sign of humanity had been swept away. Roads had vanished, and forests were fire-blackened stumps. Villages were gray smears where stone walls had crumbled together. Only the faintest outlines of the great forts of Douaumont and Vaux could be traced against the churned up background...."

The bombardment was only the beginning of Operation Judgement.

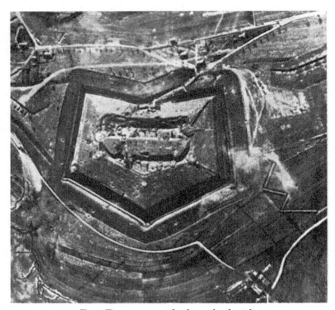

Fort Douaumont before the battle

Fort Douaumont after the battle

Following the shelling, the German infantry marched into town. Within three days, they had broken through the French defensive line and captured 10,000 French soldiers.

On February 25, the Germans took Fort Douaumont.

Now the battle was at a critical stage.

If all of Verdun fell, the Germans would then push further into France, bringing them closer to their ultimate goal of capturing Paris.

The Allies had to stop this attack—right away.

Enter General Philippe Pétain, commander of the French 2nd Army. His battle cry was, "They shall not pass!"

Pétain's first major decision was to widen that single road into Verdun. He then gathered thousands of men and supplies from the front lines and sent them into Verdun. Pétain also changed the typical service order. At that time, soldiers on the front lines served until they were seriously wounded or killed. But Pétain realized morale was sinking among his army—men despaired over the deadly stalemate on the Western Front—so he changed the Standard Operating Procedures. His soldiers would now serve fifteen days on the front lines, then transfer out.

But even two weeks at Verdun was brutal. One London newspaper report described some of the horrors men saw in this battle, including, "… peering through the moonlight at what they thought to be stealthily crawling Germans, found them to be wounded men frozen to death."

French reserves cross a river on their way to Verdun.

Then, in early March, the Germans attacked Verdun again. But this time, they faced Pétain and his reinforced French artillery forces. The Germans suffered horrendous casualties— thousands of men slaughtered. One French artillery shell landed inside a German ammunition area that was holding machine guns, bullets, and artillery rounds. When the shell exploded, it ignited all of that gunpowder, resulting in one of WWI's biggest explosions.

Crown Prince Wilhelm, an officer serving in Verdun, later described what started to go wrong for the Germans.

"... surprises were no longer possible; and the early impetuous advances by storm gave place to a gigantic wrestle and struggle for every foot of ground. Within a few weeks, I perceived clearly that it would not be feasible to break through the stubborn defense, and that our own losses would ultimately be quite out of proportion to the gains."

And yet, the fighting continued.

Now the French needed to drive the Germans out of Verdun.

Enter French General Robert Nivell.

French soldiers fire a captured German machine gun

Nivel, an artillery officer, knew how to make French guns even more deadly accurate. He launched a series of attacks and counterattacks, and by late October, his tactics were succeeding. His 200,000 French soldiers with their 700 guns won back the two Verdun forts.

German leader Paul von Hindenburg described the turn of events.

"We lost [Fort] Douaumont, and had no longer the strength to recover that field of honour of German heroism. For this attack, the French commander had abandoned the former practice of an artillery preparation extending over days or even weeks. By increasing the rate of fire of the artillery and trench-mortars to the extreme limit of capacity of material and men, only a short period of preparation had preceded the attack, which had then been launched immediately against the physically exhausted and morally shaken defenders."

The Germans had lost the element of surprise at Verdun, and now their troops were suffering from exhaustion, starvation, and the deadly fire of the French artillery. But at Verdun, the fighting continued, a constant stream of misery and destruction. Tens of thousands of men lay dead. Entire buildings had crumbled to the ground.

"The battles... exhausted our forces like an open wound," von Hindenburg wrote. "Moreover, it was obvious that in any case the enterprise had become hopeless, and that for us to persevere with it would cost us greater losses than those we were able to inflict on the enemy."

In December, the Germans changed course. After 300 days of fighting, they abandoned Verdun.

With Operation Judgement, the Germans planned to eliminate French reserves. Instead, the Battle of Verdun left 450,000 German soldiers dead or wounded. However, French losses were almost as terrible, with more than 400,000 dead or wounded.

Years later, General Pétain remembered the men who fought with him at Verdun.

"Their expressions seemed frozen by a vision of terror," he wrote, "their gait and their postures betrayed a total dejection; they sagged beneath the weight of horrifying memories."

The stalemate continued....

WHO FOUGHT?

Philippe Pétain

Henri Philippe Benoni Omer Joseph Pétain—otherwise known as Philippe Pétain—joined the French army in 1876. He served in various garrisons and showed talent on the battlefield. But he wasn't quickly promoted.

Pétain rejected the French army's reliance on infantry assaults—sending in massive numbers of men to fight a battle. Instead, he believed "firepower kills." Pétain's strategies relied on using massive artillery power against the enemy, not soldiers.

At age 58, after being told he would never become a general, Pétain bought a villa and planned to retire.

But World War I broke out, and the old officer was brought back to command French forces.

Now Pétain would prove his point that "firepower kills."

In the first five months of the Battle of Verdun, his field artillery fired over 15 million shells on the Germans. His aggressive and winning tactics earned Pétain the nickname "The Lion of Verdun."

But life has many strange turns.

Just a few years after WWI, this French hero would become a disgraced figure.

In 1940, during WWII, Hitler's Nazi army invaded France. Pétain was among a group of French military officers who agreed to work with Hitler. These Nazi sympathizers, as they were called, became part of "Vichy France"—the part of France under German rule.

After WWII, Pétain was tried and convicted of treason for betraying his country. He was sentenced to death, but due to his old age and his heroic service in WWI, Pétain's sentence was changed to life in prison.

Philippe Pétain was exiled—sent away—to a small island off the French coast.

He died in 1951. He was 95 years old.

BOOKS

The Battle of Verdun: A Captivating Guide to the Longest and Largest Battle of World War 1 That Took Place on the Western Front Between Germany and France by Captivating History

The Greatest Battles in History: The Battle of Verdun by Charles River Editors

INTERNET

Ten photos from the Battle of Verdun: time.com/4596494/battle-verdun-photos

US Military Academy map of the Battle of Verdun: www.emersonkent.com/map_archive/battle_of_verdun.htm

Battle of Jutland

May 31 – June 1, 1916

German battleship squadron

HERE'S AN INTERESTING fact about World War I: the war's last great sea battle happened in 1916—two years before the war ended.

Why no more great sea battles in the next two years?

Before we answer that question, you first need to know some things about the English and German navies.

From the year 1805 onward, England commanded the world's most powerful navy. For the next century—100 years— no other fleet could compare. England's navy kept the island nation safe and conquered territories around the world. By 1914, when WWI began, the British navy commanded

151 combat ships.

At that same time, Germany had fewer than 100 combat ships for WWI. With neither the time nor the resources to build more ships to match England's fleet, Germany decided to "level the field" by destroying some of England's fleet.

Many technological developments took place during WWI, from fighter planes and machine guns to chemical weapons. But another WWI innovation was the British ship known as the dreadnought (pronounced "dred-not.") In old English, the word "dreadnought" meant "fearless person," and these British dreadnought ships seemed fearless.

Dreadnoughts were equipped with heavy-caliber guns that had longer firing ranges and more destructive power. The guns were stationed in more turrets and came equipped with firing control systems that could help calculate where to hit an enemy's ship for maximum damage.

Two twelve-inch guns aboard *HMS Dreadnought*

The dreadnoughts also came with a new kind of engine system. It was driven by steam instead of coal. Steam provided the ships with steady high-temperature energy, whereas coal could vary greatly depending on the type and amount used. The steam drove the ships' turbines to faster speeds, giving the engine more power.

In order to protect all this new weaponry and machinery, heavy armor-plating was added to the dreadnought's decks and waterlines—where the ship's hull (sides) meets the surface of the water.

Germany had fewer ships, but many of its vessels were equipped with stronger armor and better safety features. For instance, German ships had watertight compartments. If enemy fire blew a hole in the ship, the compartment would seal itself off, allowing the ship to stay afloat for more battle or to escape.

To beat the British navy, German Vice Admiral Reinhard Scheer planned to destroy about fifty ships. These British ships were commanded by British Admiral Sir John Jellicoe. By wiping out Jellicoe's ships, the two navies would be almost equal in size—and maybe even turn Germany into the superior sea power.

German Vice Admiral Reinhard Scheer

But Scheer had another motivation.

The Allies had formed a blockade in the Atlantic Ocean—a blockade is a military maneuver that cuts off supplies, war material, and even communications from a particular area. In this case, English vessels were blocking any ships from reaching Germany that carried supplies for the Central Powers—including food. By 1916, this blockade had caused serious trouble for Germany. Its people were starving, even rioting over the lack of food.

Scheer wanted to defeat the British navy, but he also wanted this attack to break the blockade.

He came up with a clever plan. First, he would send out a German fleet—commanded by Vice-Admiral Franz Hipper—that would attack the Allied merchant ships sailing near the coast of Norway.

But this initial attack was actually a trap.

Because right behind those first German ships were even more ships—with Scheer himself onboard. When the Allied vessels least expected it, the full German fleet would attack, wiping out the English ships.

But plans are not actual battles.

Unbeknownst to Scheer, the Allies knew all about his secret plan—because the British had learned how to decipher German radio messages!

Armed with new knowledge, British Admiral Jellicoe devised a counter-plan to destroy Scheer's ships.

British Admiral John Jellicoe

Jellicoe would send a combined fleet into the fight: fifty-two ships commanded by Admiral David Beatty, and then another fleet under Jellicoe's own command, sailing down from Scotland.

Look at the maps below. On the left, you can see where the British ships left Scotland (above England) and sailed east toward Norway then turned south toward Denmark. Meanwhile, the German ships were sailing north. The battle would erupt off the coast of Denmark near an area known as Jutland—where this great sea battle gets its name.

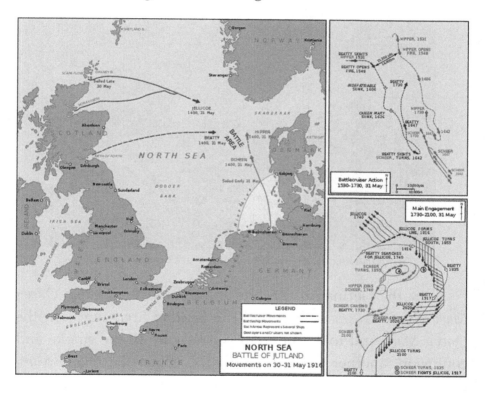

First, the German ships under Hipper tried to lure the British fleet into Sheer's trap. But both sides got drawn into battle lines—where warring ships sit parallel to each other, their broadsides exposed. The British opened fire with heavy salvos—all guns firing at once—and the German ships replied with their own fire.

Flames burst from gun turrets. Shells boomed and soared across the sea. Near-hits splashed into the ocean, throwing giant geysers of water into the air.

Then German shelling struck the *Indefatigable*. The British battlecruiser exploded—and started sinking. More than 1,000 men went down with the ship. Two sailors survived.

HMS Indefatigable sinking after being struck by German shells

Now the German battleship *Seydlitz* and the British *Queen Mary* closed in on each other. The *Seydlitz* fired twice on the *Queen Mary* and took out a rear turret gun. In the smoke and confusion, another German ship approached the *Queen Mary* and fired on her. Although a shell from the *Queen Mary* managed to destroy a small gun turret, the *Seydiltz* was still very much in the fight. It knocked out another main turret on the *Queen Mary*. Before it could lose another vessel, the British decided to sail out of gun range.

But it was too late.

The *Queen Mary* was hit two more times. One shell struck the forward magazine—where ammunition was stored—and the resulting explosion snapped the *Queen Mary* in half. The ship sank, taking down 1,266 crewmen. Eighteen men survived.

Final destruction of the battlecruiser *HMS Queen Mary*

Heavy German gunfire continued. The English were losing badly.

But suddenly the battle shifted.

The Germans had managed to capture some English sailors. Under questioning, these sailors revealed the British plan—that at least another twenty large English battleships were going to be joining this fight.

Hearing this, Scheer suddenly realized he wasn't winning this battle. In fact, Hipper's entire fleet was now at risk. And if Hipper's ships went down, the German navy would never recover. England's fleet would only grow more powerful.

The only way to save Hipper's ships, Scheer decided, was for Scheer's ships to sail into this fight, too.

But as Scheer's ships moved in for attack, flashes of gunfire opened from even more British ships—the sailors from Scotland had arrived!

Now Scheer found himself locked into the very same trap he had hoped to spring on the British.

The sea battle's second phase began.

Beatty positioned his ships to face the enemy and ordered blockade to stop any German ships from escaping this battle.

The British opened fire with a heavy salvo of shells, so heavy that Scheer soon realized his entire fleet might sink. His ships needed to get away from the British fleet. Quickly, Scheer ordered a smokescreen—his ships firing their guns to create billowing black clouds—to disguise their escape.

Germany's damaged *SMS Seydlitz* after the Battle of Jutland

As Scheer sped out of range of the British guns, Jellicoe gave chase. And soon Scheer realized he'd made another mistake.

This escape route was leading the German ships away from home port—the safest harbor.

Scheer decided to reverse course. Now his ships were headed straight toward the British fleet. Scheer hoped this sudden maneuver would split the British formation, allowing his ships to reach safety, but the British fleet held.

Once again, Scheer was facing his enemy's full guns.

But Scheer was heading in the right direction for home port, so he sent out four battlecruisers to fight a delaying action—a military maneuver that distracts the enemy while other forces escape the battlefield.

The battle stretched all day and into the evening.

As darkness fell, Jellicoe decided to stop the chase. For one thing, he was concerned about the German cruisers firing surface torpedoes. He also suspected German submarines were in the area. Both threats could sink Jellicoe's entire fleet. He decided it was better to stop now and resume fighting the next day.

But that never happened.

Under the cloak of darkness, Scheer managed to escape—and didn't look back.

The Battle of Jutland was over.

The British claimed victory. They still controlled the northern sea area, keeping the blockade in place. And for the rest of WWI, Germany never launched another effective naval challenge.

But most historians consider the Battle of Jutland a draw. Neither side really won—they simply didn't lose. The Germans had planned to destroy at least part of the English fleet and stop the blockade. They didn't. The English planned to wipe out the German navy. They didn't. Also, the British lost three battlecruisers, three cruisers, eight destroyers, and 6,784 men. Germany lost one battleship, one battlecruiser, four light cruisers, five destroyers, and about 2,500 men.

After the battle, Jellicoe was criticized for being so cautious. Some people insisted he should have pursued Sheer and destroyed the German ships. But others pointed to the high stakes at the Battle of Jutland. In fact, one of Jellicoe's strongest defenders was Winston Churchill—the man who had failed at Gallipoli.

Churchill said Jellicoe "was the only man on either side who could have lost the war in an afternoon."

WHO FOUGHT?

John Travers Cornwell, age 16, recipient of the Victoria Cross

John "Jack" Cornwell was 16 years old when he signed on to the British navy—without his parents' permission.

Jack served during WWI aboard the *HMS Chester*. His rank was "Boy, First Class."

As a sight setter, Jack was responsible for making sure the ship's guns were correctly aimed for maximum destruction. In the Battle of Jutland, Jack's ship was part of the blockade to keep German ships from escaping. When the *Chester* was scouting ahead during the battle, four German cruisers emerged from the smoke and haze. The Germans opened fire on the *Chester*. Jack was stationed at a 5.5-inch gun mounting whose shields didn't reach all the way to the ship's deck. As the shells exploded, pieces of them flew under and behind the shields, hitting Jack's gun crew and killing everyone except Jack. Despite severe injuries, the 16-year-old stayed at his post awaiting orders until the ship was pulled from the battle's action.

The entire crew of *Chester* was badly injured. As other ships sailed past it, the sailors saw limbless men laid out on the *Chester*'s deck, their arms and legs blown away by the German gunfire. And yet, the *Chester*'s men still yelled support to the other sailors. In the end, most of the Chester's crew died from their injuries.

Medics found Jack still at his gun, the sole survivor of his position. Steel shards pierced his chest. Jack was taken to a hospital, but later died from his injuries.

After his death, Jack was awarded Britain's highest award for gallantry in the face of the enemy, the Victoria Cross.

Jack Cornwell's grave has a monument that reads:

It is not wealth or ancestry
but honourable conduct and a noble disposition that
maketh men great.

BOOKS

British Dreadnought vs German Dreadnought: Jutland 1916 by Mark Stille

British Battlecruiser vs German Battlecruiser: 1914–16 by Mark Stille

World War I Seaplane and Aircraft Carriers by Mark Lardas

INTERNET

A full account of the Battle of Jutland narrated by Admiral Jellicoe's grandson. This 24-minute animation gives an overview of the major "chapters" of the battle. Graphics, animation, animated maps, and contemporary photography illustrate key points. vimeo.com/162655850

The Sea War Museum of Jutland records the dramatic stories of maritime warfare in the North Sea: www.seawarmuseum.dk/en

MOVIES

Battle of Jutland (2016) documentary commemorating the 100th anniversary of the battle

Lawrence of Arabia

August 6, 1888 – May 19, 1935

THOMAS EDWARD LAWRENCE

Photo courtesy of the National Archives of Norway

MANY HEROES CAME out of World War I—from privates to generals. But one English soldier became a legend: Lawrence of Arabia.

Born Thomas Edward Lawrence, his amazing life story inspired several Hollywood movies and hundreds of books. His WWI memoir, *Seven Pillars of Wisdom*, is still read around the world—100 years after it was published. As one WWI British general said, "There is no other man I know who could have achieved what Lawrence did."

But Lawrence's beginnings were fairly ordinary. He grew up with two brothers, graduated from Oxford College in England, and became an archeologist. In 1910, he joined an archeological dig in northern Syria, part of the Middle East. While in Syria, Lawrence learned to speak and read the Arabic language. He also came to appreciate certain parts of the Arab culture. In 1914, when WWI broke out, Lawrence was still working in Syria and volunteered for the British army.

As a lieutenant, Lawrence was assigned to the military intelligence department in Cairo, Egypt. His job included mapmaking and interviewing captured prisoners. Lawrence also wrote a pamphlet—*The Turkish Army Handbook*—to help his fellow Englishmen better understand the wartime culture of Arabic people.

But office work didn't satisfy Lawrence's passion for adventure, and when he heard his two brothers had been killed in battle, Lawrence asked to join the front lines. He also wanted to help the Arab people who were fighting against the Ottoman Empire.

You've already read about the rise and fall of the Ottoman Empire. By the early 1900s, a good number of Arabs living inside the Empire were rebelling against the Turks. These rebel

Arabs wanted independence for their countries. To gain it, they fought alongside the Allies—who also wanted to defeat the Turks.

Ottoman Camel Corps in the Middle East during World War I

In 1916, the army promoted Lawrence to captain and sent him on a secret mission. The Ottoman's Turkish Army had trapped some British and eastern Indian soldiers in an isolated location. The fate of these Allied soldiers looked grim. But Lawrence managed to get all the wounded soldiers released. However, the Turks took the other soldiers as prisoners.

This first mission wasn't a total success, but Lawrence wrote a report about it that was so insightful his commanders sent him on more missions.

With each day Lawrence spent in the Middle East, he gained more and more knowledge about the Ottoman Turks and the rebel Arabs. Lawrence passed this information onto the British military and the Allies. He had a knack for understanding subtle but crucial differences between the cultures. For instance, Lawrence advised the British soldiers:

"Do not try to do too much with your own hands. Better the

Arabs do it tolerably than that you do it perfectly. It is their war, and you are to help them, not to win it for them. Actually, also, under the very odd conditions of Arabia, your practical work will not be as good as, perhaps, you think it is."

T.E. Lawrence, 1918

Some of the Arab tribes fighting against the Ottomans were known as Bedouins (pronounced: bed-oh-ins"). A nomadic people who wandered the desert, the Bedouins never settled down permanently in one place. In warfare, they showed both good and bad qualities. On the good side, the Bedouins were ferocious fighters, especially when their honor was at stake or if they could reap great rewards from the battle. On the bad side, the tribes weren't organized like a trained army. If an upcoming battle was taking too long to start, the Bedouins might just pack up their tents and ride away on their camels.

But the biggest problem among the rebel Arabs was the feuding—the fights simmering among the different clans and tribes. Some feuds went back generations, and some still continue to this day!

In 1916, when the Arab revolt broke out in full force, the British government hoped to find leaders within the rebellion who could help the Allies defeat the Turks. Lawrence met with the rebellion's leader, King Hussein bin Ali. The king had four sons. Lawrence decided the strongest leader among them was Hussein's third son, Prince Emir Faisal. The king agreed with Lawrence.

Prince Emir Faisal

As the revolt spread, Lawrence was the main link between the British government and the Arabs who were fighting the Ottoman Empire.

And that is where the true legend of Lawrence of Arabia began.

At first, Lawrence simply watched and listened while the prince handled different situations among the tribes. For instance, Faisal paid enormous amounts of gold to the Bedouins, keeping them available if he needed them to fight. The prince also made use of his deep knowledge of the clans to negotiate peace among them. Faisal's leadership technique was to remain calm yet clever, never resorting to anger or impulsive actions. In turn, Faisal observed Lawrence and soon realized this British officer and archeologist respected the Arab people and would make a trustworthy partner. To show his appreciation, Faisal gave Lawrence many gifts—including white and gold Arab robes.

Lawrence in Arabia, 1917

These robes served two purposes, Lawrence realized.

"Feisal asked me if I would wear Arab clothes like his own while in the camp," Lawrence later wrote. "I should find it better for my own part, since it was a comfortable dress in which to live Arab-fashion as we must do. Besides, the tribesmen would then understand how to take me... If I wore [Arab] clothes, they would behave to me as though I were really one of the leaders."

Soon Lawrence was leading the tribes into battle. Imagine the sight of this Englishman riding on an enormous camel, his white-and-gold robes flashing in the bright sunlight as he raised his pistol and yelled, "Charge!" At Lawrence's command, hundreds of Arabs on horseback and camel would emit blood-curdling screams and descend on the Ottoman. Hearing the tribal screams, the Turks became almost paralyzed with fear.

Arab rebels very rarely took prisoners—their screams meant the end was near.

Lawrence didn't have any battlefield training from the British army. And at first, he felt this rebellion was too bloody. But by the end of it, Lawrence himself was ordering entire massacres.

Lawrence of Arabia's reputation as a warrior spread throughout the Middle East and beyond. From 1917–18, he fought in many battles.

Here are some of his most famous clashes.

The Battle of Akaba

July 6, 1917

Lawrence at Akaba

AKABA WAS A seaport in the country of Jordan. The Turks had stationed only about 750 soldiers there. Compared to the powerful British navy, it seemed like the Allies could take control of Akaba.

But the channel of water leading into Akaba was narrow, and the Turks had their guns facing the sea (kind of like they did at the Dardanelles in the Battle of Gallipoli). Eventually, the British did manage to land some marines near Akaba, but the overall mission was a failure (also like Gallipoli!).

It seemed like nobody using the waterway could drive out the Turks from Akaba. And the backside of the port wasn't any easier. It was just a vast desert—about 200 miles of nothing but sand—and nobody was foolhardy enough to try crossing that desert....

Except Lawrence of Arabia.

Lawrence wanted to ride with the Arabs across that huge stretch of hot sand and attack the Turks from behind. Prince Faisal didn't agree with this plan. But Lawrence finally convinced him to try it.

Lawrence and approximately 2,000 Arab rebels risked crossing the barren desert. When they finally reached the outskirts of Akaba, a fierce battle against the Turks broke out—guns blazing, swords flashing. Lawrence and the Arabs took the high ground above a blockhouse where the Turks were hiding. The Arabs started firing their rifles. These guns, made in 1890, were considered less than great. Meanwhile, the Turks were armed with modern rifles and artillery that should've wiped out Lawrence and the Arabs.

Instead, Lawrence's men fired their outdated weapons so fast and frequently that their hands blistered from the firepower's heat. Clever Lawrence had also positioned his forces so that the sun was shining in the Turks' eyes—they couldn't find clear targets!

The Arab rebels won the battle, took over the blockhouse, and killed many Turks.

Now Lawrence and his forces approached the city of Akaba itself. At the same time, the British navy opened fire on the city's port. It was a surprise attack from both forward and back, and it forced the Turks to surrender Akaba to the British.

While Akaba wasn't an extensive battle, it was a positive change. Lawrence had led Arab ground forces to victory, proving to the British commanders that these rebel tribes really wanted to topple the Ottoman Turks. All they needed were better weapons.

Attacks on the Hejaz

Lawrence with his Arab dagger

AFTER THE SUCCESS of Akaba, the British started arming the Arabs with modern weapons—newer rifles, machine guns, mortars, and even armored cars.

The Ottoman Turks were using an extensive railroad system to supply their army. The rail system was called the Hejaz, with routes traveling throughout the desert landscape.

Lawrence now set his sights on destroying that railway system.

Armed with their modern weapons, his men laid mines and dynamite along the iron tracks. As a train approached, the Arabs would detonate the explosives and blow up the iron rails, forcing the train to stop. Then Lawrence's rebel force—screaming its war cry—would descend on the trains that carried Turks inside.

"We had a Lewis [machine gun] and flung bullets through the sides," Lawrence wrote to a friend describing one such attack. "So [the Turks] hopped out [of the train] and took cover behind the embankment, and shot at us between the wheels at 50 yards."

Any Turk who tried to escape was cut down by Arab rifle and machine-gun fire, or he was sliced to death by Arabs racing forward on camels, swinging their swords.

WWI Arab cavalry

By late October 1918, Lawrence was promoted to full colonel. His forces had the Ottoman Army in full retreat with soldiers surrendering to the Allies in droves. By November 1918, when WWI ended, the Ottoman Empire was almost completely demolished.

After the war, Lawrence returned to England. He was awarded many honors for his wartime leadership and bravery. But he struggled in his life away from Arabia. In 1922, he joined the Royal Air Force (RAF) under the name John Ross. But when his true identity was discovered, he was thrown out of the RAF. Lawrence then joined the royal tank force under another name, only to be thrown out again.

T.E. Lawrence on his motorcycle

Lawrence even worked for Winston Churchill for awhile. He also wrote several more books. But the war seemed to have made him restless and almost tortured by what he'd experienced. Lawrence was also unhappy with the British government because it didn't keep its promise to the Arab people. Instead of granting the Arab rebels their independence, the Allies divided up the Middle East among themselves. Lawrence considered it a terrible betrayal—so much of a betrayal that he refused to be knighted by the King of England.

Eventually, Lawrence turned his passion toward motorcycles.

On May 13, 1935, he was speeding down an English country road when he swerved to avoid hitting two boys on bicycles. The motorcycle crashed. Lawrence was thrown over the handlebars and suffered horrific head injuries.

Thomas E. Lawrence later died of those injuries. He was 46 years old.

BOOKS

Sterling Point Books: Lawrence of Arabia by Alistair MacLean, audio narrated by Peter Ganim

Lawrence of Arabia: The Life and Lega.cy of T.E. Lawrence by Charles River Editors

The Sinai and Palestine Campaign of World War I: The History and Legacy of the British Empire's Victory Over the Ottoman Empire in the Middle East by Charles River Editors

INTERNET

Lawrence of Arabia gained great wisdom during his lifetime. Among his quotes:

"It is difficult to keep quiet when everything is being done wrong, but the less you lose your temper the greater your advantage."

More of his quotes can be found here: www.azquotes.com/author/8582-T_E_Lawrence

Want to see authentic footage of Lawrence of Arabia and Prince Faisal in action? Their military campaigns in the desert? Internet Archive has some filmed scenes available here: archive.org/details/34690HDWithLawrenceofArabiaMos18fps

MOVIE

Nothing compares to David Lean's captivating biopic, *Lawrence of Arabia.*

Here is a short trailer of the movie:

youtube.com/watch?v=qPQ7CR3wn8A

Battle of the Somme

July 1, 1916 – November 19, 1916

Scottish Highland regiment advancing under fire at the Somme, 1916

NOW LET'S RETURN to the Western Front in Europe.

By the summer of 1916, the Allies and the Central Powers were both now running low on ammunition, food, and other supplies. More importantly, both sides were losing soldiers at an alarming rate. Every week, trench warfare wiped out thousands of men. But still the stalemate on the Western Front dragged on.

Both sides needed the stalemate to end.

Unfortunately, ending the stalemate created one of the bloodiest clashes in military history: the Battle of the Somme.

Allied troops marching towards the Battle of the Somme

Also known as the Somme Offensive, this battle wound or killed 60,000 soldiers—on its first day.

The Somme River was part of the Western Front. In that area, the Germans held the high ground—or, the strongest defensive position—and they had about one million soldiers in the area.

To defeat them, France sent more than one million soldiers to the Somme. The British dispatched another 1,500,000 soldiers, giving the Allies about a two-to-one advantage here over the Central Powers.

This numerical advantage was important. Up until this time in history, a basic truism of warfare was that if one army held the stronger defensive position—here, the Germans—then the other side would need about a three-to-one advantage in manpower to defeat them. Another truism was that the army with the most men usually won the battle.

However, WWI's modern warfare tactics were about to change those two ideas.

Look at the map of the Western Front. France sits at the map's bottom, Belgium is at the top (also notice the southern tip of England, across the water). Germany is to the right side of the map. And finally, that line running among these countries marks the Western Front. Now look at the two shaded "bulges" in that line. The lower bulge represents the Battle of Verdun, which you already read about. The upper bulge marks the Battle of the Somme.

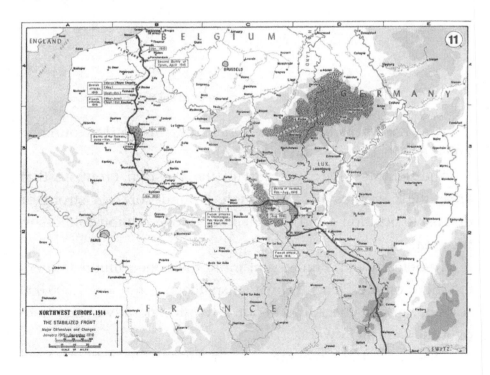

Just before this battle broke out, the British government had passed a new law. The country was running low on soldiers, and now the government was going to expand the draft. Married men, for instance, could now be included in the armed services. These fresh recruits would be among the first Allied soldiers sent to the Somme.

The Allied plan for the Somme was to dispatch a massive number of soldiers, a force so large it would overwhelm and ultimately defeat the Germans, thus breaking the stalemate on the Western Front. The Allies could then push the Germans out of France and gain the offensive advantage of the war. That momentum would take some pressure off the Russians struggling against the Germans on the Eastern Front. The officer given command of the Somme Offensive was British Field Marshall Douglas Haig.

In late June 1916, the Allies bombarded twenty-five miles of the German line near the Somme River. Within the first five days alone, the Allies fired more than one million artillery rounds.

On July 1, the Allies launched an infantry attack.

But early on, this operation started to fall apart.

The initial five-day bombardment mostly served to warn the Germans that another attack was coming. Just as troubling, the artillery shelling didn't destroy the German defenses—in fact, many of the shells were defective and didn't even explode! The shells that did explode ripped deep holes in the ground, leaving rough terrain that the Allied soldiers now needed to run across. And just as bad, the bombardment dispersed— spread out—the barbed wire the could slit open men's bodies.

German infantry

Another problem for the Somme Offensive was the timing. British commanders wanted to launch this infantry attack at first light—around 5:00 a.m.—but French commanders insisted on waiting until 7:30 a.m.

The element of surprise was lost.

When Allied commanders blew their whistles, the soldiers—remember, many new recruits in the British draft—climbed from their trenches and faced a chewed-up landscape full of barbed wire. On their backs, these soldiers carried backpacks that weighed about 60 pounds.

Why were the backpacks so heavy? Because here's what was inside them: tent parts and tent pole, ammunition, first aid kit, canteen and its cup, baking tin, bread rations, blanket, towel, soap dish, shaving kit, handkerchief, foot powder, and extra socks. But that wasn't all! Strapped to the outside of the pack were the soldier's bayonet, shovel, trench-digging tool, and mess kit—with cooking and eating utensils.

Trench warfare had many deadly aspects, but among the worst hazards was entering no man's land where enemy trenches were aimed to kill. But making matters worse for the Allied soldiers were these heavy packs that slowed them down and gave the Germans more time to shoot at them.

One British soldier later described what it felt like to race from a trench into no man's land.

"Then five minutes to go. And then zero hour, and all hell lets loose. There's our barrage, the Germans' barrage, and over the top we go. As soon as you get over the top, fear has left you, and its terror. You don't ... look, you see. You don't hear, you listen. Your nose is filled with fumes and death....The veneer of civilization has dropped away."

Canadian soldiers fighting at the Somme

The slaughter at the Somme was immediate.

That German artillery that had survived the initial bombardment now mowed down the Allied soldiers who were staggering under the weight of their backpacks. In mere minutes, the Allies lost more than 80 percent of their force!

By the end of the first day, 60,000 soldiers were wounded or dead.

Allied soldiers captured by the German Army

After these catastrophic casualties on the first day, some people thought that Allied commander Haig should have pulled out the troops.

But the Battle of the Somme raged on—for months!

By September, dead bodies smothered the battlefield. Some men were so disfigured they could not be identified.

And still the whistles blew.

Still the men went over the top.

Still they raced into no man's land.

"The trench was a horrible sight," one British captain later said. "The dead were stretched out on one side, one on top of each other six feet high. I thought at the time I should never get the peculiar disgusting smell of the vapour of warm human blood heated by the sun out of my nostrils. I would rather have smelt gas a hundred times."

The British people back home didn't know about this slaughter at the Somme because the military was sending back false casualty reports. But as families received news about their loved one's death, the horrific truth leaked out. People were outraged.

Now Allied commanders needed drastic measures to end this battle.

Enter the Steel Monster.

British WWI tank

Tanks grew out of the same idea that invented farm tractors capable of rolling across rough fields—why not have something like that on the battlefield, too? With their enclosed metal shields that protected the men inside, tanks could also repel bullets and shells while the rotating tracks propelled them across rough terrain into enemy territories.

In September 1916, armored tanks made their first appearance at the Somme. Although a new invention, tanks soon became a signature weapon of WWI.

At first, the tanks caught the Germans off-guard. Not many weapons could combat these gigantic machines of death. Unfortunately, these first British tanks were slow, with a top speed of about 3 mph, and not very maneuverable. They also broke down a lot—the British launched almost fifty Mark I tanks at the Somme, but only about half made it to the front lines.

English tank with its tread blown off

Even with the tanks, the Battle of the Somme fell into another stalemate of bloodshed and death.

Then winter arrived.

Freezing wind.

Cold rain.

Darkness.

The trenches turned into frozen swamps. After all the months of deadly fighting, Allied soldiers felt little desire to continue. As one Somme soldier wrote in a letter to home:

"It's the end of the 1916 winter and the conditions are almost unbelievable. We live in a world of Somme mud. We sleep in it, work in it, fight in it, wade in it and many of us die in it. We see it, feel it, eat it and curse it, but we can't escape it, not even by dying."

Hoping to win, the Allies made one last push for an important ridge held by the Germans. The Allies managed to take it, but the rest of the massive battlefield remained a vast wasteland of destruction and more death.

Despair haunted both sides.

"The tragedy of the Somme battle," said one German soldier, "was that the best soldiers, the stoutest-hearted men were lost; their numbers were replaceable, their spiritual worth never could be."

After five months of fighting, the Allies had advanced only seven miles from where they started, but more than 600,000 men were dead.

Finally, in November 1916, the Allies called off the Battle of the Somme.

After 141 days of fighting, there were more than one million casualties—and the Western Front was still a stalemate.

"Somme," complained one German officer. "The whole history of the world cannot contain a more ghastly word."

WHO FOUGHT?

British Field Marshal Douglas Haig

A senior officer of the British Army and commander of the British Expeditionary Force (BEF) on the Western Front, Field Marshall Douglas Haig earned two very different nicknames after the Battle of the Somme: "Master of the Field" and "Butcher of the Somme."

Which nickname was right?

That depends on how you view the Battle of the Somme and WWI in general.

Immediately after the war, Haig's reputation remained high. He had led the British during the final Hundred Days Offensive when Allied soldiers broke through the German line and captured 195,000 German prisoners. Haig's aggressive campaign helped end WWI and is considered by many to be among the greatest victories ever achieved by a British-led army.

In 1921, Haig's funeral was declared a day of national

mourning.

But later, around the 1960s, some historians began to criticize Haig's choices during WWI. They labeled him a "butcher" for the two million British casualties that happened under his command.

Go read more about Haig, and decide for yourself which nickname is correct—or whether both are.

BOOKS

Great Push The Battle of the Somme 1916: Photographs from Wartime Archives by William Langford

Armoured Warfare in the First World War (Images Of War) by Anthony Tucker-Jones

The Germans on the Somme (Images of War) by David Bilton

INTERNET

The British Broadcasting Company (BBC) has some interesting quizzes about the Battle of the Somme. See if you can pass the test!

There's also a time-lapse interactive map showing the battle, hour by hour. bbc.co.uk/guides/zy98xsg

History.com's video on the Battle of the Somme includes some historic footage: www.history.com/topics/world-war-i/1916-battle-of-the-somme-video

MOVIES

They Shall Not Grow Old (2018)

Director Peter Jackson has created an unforgettable experience by combining actual WWI footage with the recorded remembrances of soldiers who served in the battles.

A Turning Point

1917

WWI Russian troops awaiting German attack, 1917

BEFORE WE GET to our next battle, you need to know about two critical events. Both happened in 1917. And both had significant impact on the war.

First, the Russian Revolution.

Russia was ruled by a royal Czar—sort of like a king. For hundreds of years, Russian Czars had lived in grand palaces eating fancy meals and throwing expensive parties. But by the early 1900s, ordinary Russians were very poor. The country wasn't well managed, and many people were starving—even while the Czar's parties became even more lavish. By 1917, some new political leaders in Russia were demanding an end to the Czar—even if that meant killing him.

Russian women marching for "bread and peace," 1917

The political and cultural turmoil inside Russia forced the Czar to pull his army out of WWI. Not only was the war unpopular with most Russians, but the Czar also needed his soldiers back home to quell a civil war—that's a war happening within one country, fought by different groups trying to gain control.

But Russia's problems turned into Germany's advantage.

When Russia exited WWI, Germany no longer had to fight a two-front war. All those German soldiers on the Eastern Front could now move to the Western Front.

That shift of soldiers spelled doom for the Allies.

But another big change was coming.

The United States was joining the Allied forces.

Why did the US join now—when WWI was already three years old?

Many reasons. But here are two important factors.

In February 1917, British and American intelligence agencies intercepted a secret German message being sent to Mexico. The note was called the Zimmerman telegram. In it, Germany told Mexico that if the US joined the Allies, Mexico should join the Central Powers. That way, when the Central Powers won the war, Germany would give Mexico parts of the United States, such as the states of California and New Mexico.

The second factor was Germany changed its military policy toward American ships.

During the war's early years, Germany had allowed neutral ships—vessels belonging to countries not taking sides in this war—to reach their destinations. But with the Allied naval blockade in the North Sea, the Central Powers were suffering. Military supplies weren't getting through to the troops, and in Germany, people were rioting over food shortages.

Germany wanted to break the blockade. In early 1917, it launched unrestricted submarine warfare. That meant *any* ship from *any* country that didn't support the Central Powers would be sunk.

Look at the map. The shaded parts mark the areas where German submarines patrolled on unrestricted warfare.

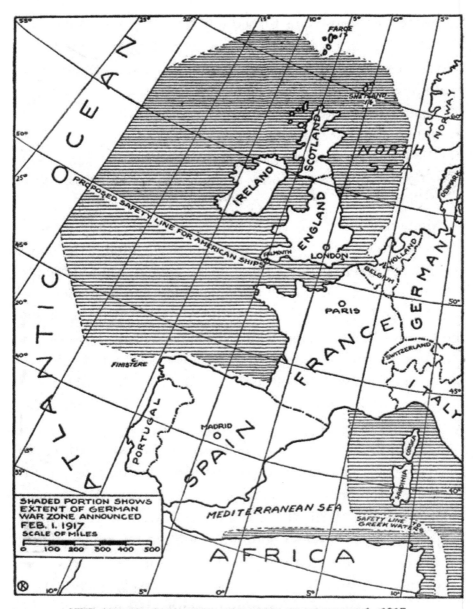

NEW GERMAN SUBMARINE WAR ZONE OF FEBRUARY 1, 1917

The German submarines, called U-boats, started torpedoing American merchant ships—vessels that were not official navy ships. In March 1917, just after that Zimmerman telegram was intercepted, German torpedoes sank four American merchant ships.

The aggression demanded an American response.

On April 6, 1917, the United States declared war on Germany.

At that time, America had a standing army of only 126,000 volunteers—not nearly enough soldiers to win victory overseas. So in May 1917, Congress passed the Selective Service Act, otherwise known as "the draft."

Young men registering for the draft, New York City, June 5, 1917

The draft required every male between the ages of 21 and 31 to register for armed services. The men who passed an initial health exam then received military training. But the time needed for that training meant most American soldiers wouldn't reach the Western Front for several more months.

Germany realized it needed to strike the Allies—hard—before the American soldiers arrived.

As you know, the Allies were struggling through the stalemate on the Western Front. But in early 1917, before the Americans arrived, things got worse.

In April, the French Army attacked the Germans. France lost 180,000 men—dead or wounded—and the survivors faced still more misery. Cold spring rains had soaked the marshy ground on the Western Front, leaving soldiers in constant water and mud. The chilly, wet conditions contributed to many diseases. Deadly pneumonia was one. Another was "trench foot"—where the skin on men's feet rotted so severely that toes and feet had to be amputated—and "trench mouth"—when bacteria-infected gums blistered and bled.

Even more health problems arose from the boxes in the trenches that were used as latrines—toilets. When rain fell, the boxes flooded and spilled their unsanitary human waste. Rats soon invaded the trenches, bringing fleas and lice that carried contagious diseases. The rats also chewed on men's bodies while they slept—and as they dying.

On top of these horrible conditions, WWI soldiers endured near-constant gunfire, and after three years of war, soldiers began developing "shell shock." Today the condition would be called Post Traumatic Stress Disorder, or PTSD.

Under these conditions, with no clear outcome for victory in sight, French soldiers began to mutiny—refusing to obey orders. Other soldiers deserted—running away. To regain control, French officers made an example of forty-three deserters. The men were lined up and killed by a firing squad. The lesson was clear: Any soldier who disobeyed orders or fled the battlefield would be shot.

Meanwhile, amid all this trouble, British General Douglas Haig decided it was time for his army to get more aggressive.

Haig wanted the British to attack the Germans in the Flanders region of Belgium.

Look at the map. The darkest line shows the Western front. At the top, find the first bulge in the Front. That's where Haig wanted to attack. Right above that, you'll see the North Sea (Straight of Dover), where German U-boats were operating. England is across the water.

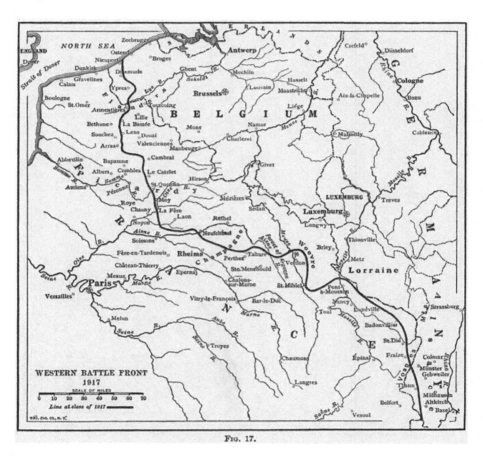

FIG. 17.

Haig believed if British forces could break through the German lines at this location, they could march for the English Channel and liberate the ports controlled by the Central Powers. Once the Allies got the ports back, they could then destroy the submarine bases used by the German U-boats.

Haig's aggressive plan was named the Spring Offensive.

If this plan had succeeded, WWI might've ended more quickly.

Instead, the Spring Offensive turned into one of history's greatest military disasters—the Battle of Passchendaele.

And that's our next battle.

The Battle of Passchendaele

July – November 1917

Australian field artillery soldiers crossing a duckboard track
passing through Chateau Wood.

LET'S GO OVER Haig's plan for this battle.

The plan had three stages, with specific tactics and strategies for each stage.

First, the Allies would take the high ground at a place known as Messines Ridge. That ridge was south of the town of Ypres, Belgium (pronounced "ee-pruh."). Second, the main attack would launch east of Ypres and capture the heights of another Belgian ridge along with the village of Passchendaele (pronounced "pash-en-dale."). This position would give the Allies the advantage of more high ground overlooking the Flanders region.

Lastly, the Allies would push toward the coast, take control of the ports, and destroy the German U-boat operations.

French commanders didn't think Haig's plan would succeed. For one thing, even if the British forces captured the submarine operations, the Germans still had other places to launch U-boats.

But Haig pressed forward, insisting this campaign could be completed in eight days.

British anti-aircraft gun on the Western Front

In late July 1917, Haig ordered a preliminary bombardment on the German forces near Ypres. The Allies blasted the Germans for two weeks. About 1,000 guns fired more than four million shells!

Then came the main attack.

Twelve British divisions and two French divisions—around 100,000 men—moved forward. Unfortunately, the artillery barrage had warned the Germans that an infantry attack was coming.

Now the Germans were prepared for it.

Another problem for the Allies was the Flanders region itself. The wet and marshy ground was full of soft clay soil. Allied tank commanders had warned Haig that their machines would struggle in the heavy mud. Haig ignored their concerns.

AUSTRALIAN WAR MEMORIAL   E00963

A supply mule team struggles in the mud

On June 7, the battle of Messines Ridge opened. Despite the bombardment, this part of the attack was still a surprise because the British had dug tunnels under the German lines and packed the caverns with explosives. Early one morning, they lit the fuse. At 3:10 a.m., the explosion went off, blowing a hole in the ground that spread 300 feet across and 70-feet deep—that's about the same height as a seven-story building!

The explosion killed many German soldiers, buried others, and left still more too dazed and confused to organize a counterattack.

The British overtook the German front-line trenches.

But the Germans held them there, and a week of fighting ensued. The battle cost the British 24,000 men, wounded or dead. The Germans lost about 20,000 men.

Despite the heavy losses, Haig moved to his second phase, the Battle of Passchendaele.

But this attack wasn't any surprise. Captured British prisoners told the Germans about Haig's plan. Even some English newspapers got ahold of the plan and published it.

German leaders now ordered more troops to the front lines and reinforced their defenses, adding more barbed wire and concrete pillboxes. A pillbox is a small fortification, usually round like a pill, with enough room inside to hold machine gunners or riflemen.

Even without this reinforcement, the German artillery was already inflicting 500 casualties—dead or wounded—each day on the British.

Now they planned to do even more damage.

Australian wounded on the Menin Road, 1917

On July 18, the British artillery opened fire on the German lines. Then about 100,000 British soldiers moved out in a creeping barrage—that's when the artillery fires ahead of the infantry to remove obstructions such as barbed wire or machine-gun nests, allowing the ground troops to move forward more safely.

Unfortunately, when British soldiers raced out of their trenches on July 31 at 3:50 a.m., the artillery wasn't much help.

Instead, the artillery shells had sunk into the mud. Some didn't even explode. The German pillboxes were almost untouched and the barbed wire remained intact. And the Allies' creeping barrage was being fired too far ahead of the infantry, leaving British soldiers to fight every enemy gun emplacement on their own.

Allied soldiers knee-deep in mud rescuing the wounded.

One Allied private later described the scene: "As far as the eye could see was a mass of black mud with shell holes filled with water. Here and there broken duckboards, partly submerged in the quagmire; here and there a horse's carcass sticking out of the water; here and there a corpse. The only sign of life was a rat or two swimming about to find food and a patch of ground. At night a yellow mist hung over the mud; the stench was almost unbearable. When gas shells came over the mist turned to brown."

Just as the tank commanders had predicted, the machines got stuck in the mud. The colossal steel monsters turned into easy targets for German artillery.

When the Germans counterattacked, they pushed the British back.

At night, rain fell, further swamping the marshy ground.

"The low lying [clay] soil torn by shells and sodden by rain turned into a succession of vast muddy pools," Haig later wrote. "The valleys of the choked and overflowing streams were speedily transformed into long stretches of bog, impassable except by a few well-defined tracks, which became marks for the enemy artillery. To leave these tracks was to risk death by drowning, and in the course of the subsequent fighting on several occasions both men and pack animals were lost in this way."

British troops advanced only two miles—at the cost of 30,000 men.

By August, British and French casualties had climbed to 74,000.

The Germans had lost 50,000 men.

But now the battle would shift, due to the appearance of General Sir Herbert Plumer and the British Second Army.

Soldiers going over the top from the trenches

Plumer wanted to try something new.

Instead of linear infantry lines that were spread across an open field, Plumer organized his troops in a tight, close formation. This idea had its flaws, too, because WWI machine gun could fire about 600 rounds a minute. A concentrated formation gave the enemy a better target.

By mid-September, another 20,000 Allied soldiers were lost—for about a one-mile advance.

The Germans counterattacked again, making full use of chemical gas that burned and blinded the Allied soldiers. An Australian division attacked the Germans and gained one mile, but it cost them 15,000 men.

Rain continued to soak the ground. Artillery shells sank into the mud, sometimes as deep as three feet. Field conditions became so treacherous that sometimes it took fourteen hours to bring one wounded man back to a safe area.

The Germans didn't have it any easier at Passchendaele. One newspaper reporter near the front lines wrote:

"The suffering of all the German troops, huddled together in exposed places, must be as hideous as anything in the agony of mankind, slashed to bits by storms of shells and urged forward to counterattacks which they know will be their death."

But in early November, the 1st and 2nd Canadian Divisions captured all of Passchendaele Ridge and the village, too.

On November 10, the Battle of Passchendaele officially ended.

But there was no attack on the coastal ports.

Haig's eight-day plan had stretched beyond three months.

The battle's final costs were staggering.

British forces suffered 244,897 casualties. Some 90,000 men were reported missing—captured, dead, or deserted. Almost half of those 90,000 men were never found. Some could have drowned or were buried in the mud. To this day, farmers tilling the soil in the Flanders region of Belgium continue to uncover the bones of WWI soldiers.

The Battle of Passchendaele is now remembered as one of the worst examples of WWI fighting. Brutal clashes. Horrific conditions. Monumental death tolls. Senior leaders who showed little regard for the men dying under their command. And all of that for very little gain.

On the Western Front, the stalemate dragged on.

In his 1938 memoirs, former British Prime Minister Lloyd George wrote, "Passchendaele was indeed one of the greatest disasters of the war.... No soldier of any intelligence now defends this senseless campaign."

WHO FOUGHT?

Frank McKinnon standing with his girlfriend, Pearl.

After his mother died of tuberculosis and his father was buried alive in a workplace accident, Frank McKinnon became an orphan.

By age eighteen, McKinnon was supporting himself by working in a Canadian munitions factory that made artillery shells. The day after turning age nineteen, McKinnon enlisted in the Canadian Army. As a private, he earned about $1 a day.

McKinnon became part of the 102nd infantry battalion of the Canadian Army. His battalion fought at the Western Front, including in the Battle of Passchendaele.

After the Canadian troops captured the village of Passchendaele, McKinnon's battalion was ordered to support an area near the front line. Carrying stretchers, soldiers searched the muddy battlefield for the dead and wounded. All the while, the Germans continued to shell them.

On November 16, McKinnon's battalion was sent to the front lines to relieve another unit. Although the Battle of Passchendaele was officially over, skirmishes and shelling continued. McKinnon was in a trench when a German shell exploded inside and a piece of shell's casing tore into his head. He was instantly killed, joining the nearly 16,000 Canadians killed at Passchendaele.

McKinnon's hometown newspaper announced his death with just a few lines.

"He had been in France but for two months before being killed," the Brantford Expositor said.

Frank McKinnon is believed to be the last Canadian killed in the Battle of Passchendaele.

BOOKS

The Battlefields of the First World War: From the First Battle of Ypres to Passchendaele by Peter Barton and Peter Doyle

Passchendaele – Ypres: The Fight for the Village (Battleground Europe Series) by Nigel Cave

Great Battles of World War I: Stunning 3-dimensional computer graphics recreate the most important battles of World War I, from Passchendaele to the Argonne by Anthony Livesey

INTERNET

The Battle of Passchendaele, a documentary marking the 100th anniversary of The Great War: youtube.com/watch?v=dBULkD_FEnw

DK Books has a website with some really good photos and information. You can search for anything WWI-related: www.dkfindout.com/uk/history/world-war-i

Sergeant York

December 13, 1887 – September 2, 1964

Battle scene of Alvin C. York, painted in 1919 by artist Frank Schoonover

YOU'VE READ ABOUT the most world-famous soldier of WWI, Lawrence of Arabia. But within the United States, another man was even more celebrated. He was known as Sergeant York.

Alvin Cullum York began the war as a "conscientious objector." That's someone who refuses to serve in the armed forces or carry a weapon because of religious or moral beliefs.

Amazingly, this same Alvin York later became one of the most highly decorated soldiers of WWI.

Born December 13, 1887, Alvin grew up with his 10 siblings in a two-room log cabin in Pall Mall, Tennessee. Alvin was a great shot from a young age. He hunted squirrels, raccoons, birds, wild boars, and deer with a muzzle-loading rifle—a rifle that's loaded by pushing a single shot down the gun's barrel.

The York family was so poor they didn't have money or time for education. Alvin dropped out of school in the third grade. During Alvin's teen years, his father died. Alvin soon turned into a "hell-raiser," drinking in saloons and getting into fistfights. But after his best friend was killed a bar fight, Alvin realized he could be next. He decided to change his ways. After attending a Christian revival, Alvin devoted his life to the teachings of the Bible.

In 1917, the new US military draft required Alvin to sign up for service.

Here is a copy of Alvin York's draft card. Look at Line 12, which asks, "Do you claim exemption from draft (specify ground)?"

"Yes," York wrote. "Don't want to fight."

The draft board denied his request.

Alvin York was sent to Camp Gordon, Georgia for basic training. He became a member of Company G in the 328th Infantry attached to the 82nd Division, known as the "All American Division."

Among his fellow soldiers, York seemed like a strange mix of characteristics. He was a conscientious observer who didn't want to fire a gun—and he was the best marksmen, able to hit almost any target with one shot. During boot camp, York agonized about being a soldier. Finally, his company commander convinced him that sometimes God needed men to fight wars to combat evil. That reasoning made sense to York.

In 1918, Private York was sent to France. His company would fight on the Western Front in an area known as the Meuse-Argonne (pronounced "muse-ar-gon"). York was soon promoted to corporal.

On October 8, 1918, Corporal York and sixteen other soldiers were sent out before dawn to take control of a nearby railroad. But their map was printed in French, which none of them spoke well and soon York and his men got lost—and wound up behind enemy lines!

York in uniform, 1919

When some German soldiers discovered York and his men, a firefight broke out. The Americans won the skirmish and took prisoners. But above them on a nearby hill, some German machine gunners realized there were only seventeen American soldiers—not an entire company. Hollering to their fellow Germans to lay down, the gunners opened fire. Everyone ran for cover.

Except Corporal York.

"I sat right where I was, and it seemed to me that every machine gun the Germans had was shooting at me," York later said. "All this time, though, I was using my rifle, and they was beginning to feel the effect of it, because I was shootin' pretty good."

York took out eighteen Germans—with eighteen shots.

He then ordered his fellow Americans to guard the German prisoners while York ran to attack the rest of the German position, including the machine-gun nest. When his rifle ran out of ammunition, six Germans charged at him with their bayonets. York pulled out his pistol and killed them all.

After this attack, York was promoted to Sergeant. Later he was awarded nearly 50 military medals, including honors from France, Italy, and Montenegro. The United States presented him with its highest military award, the Medal of Honor.

After WWI, York returned home to Tennessee. He was now a famous American hero, but he shied away from publicity. It wasn't until many years later that York agreed to a movie based on his life.

Sergeant York debuted in 1941. It sold more tickets than any other film that year.

In one scene, York explained his heroic battlefield actions to his superior, Major Buxton.

> **York**: Well I'm as much against killin' as ever, sir. But … when I hear them machine guns a-goin', and all them fellas are droppin' around me… I figured them guns was killin' hundreds, maybe thousands, and there weren't nothin' anybody could do, but to stop them guns. And that's what I done.
>
> **Maj. Buxton**: Do you mean to tell me that you did it to save lives?
>
> **York**: Yes sir, that was why.

Maj. Buxton: Well, York, what you've just told me is the most extraordinary thing of all.

York spent much of his post-war life raising money for early education. He opened the Alvin C. York Institute to help children who, like himself, might not get an education unless someone helped them pay for it.

In 1964, Alvin York passed away.

Hearing of his death, President Lyndon Johnson said, "As the citizen-soldier hero of the American Expeditionary Forces, [Sergeant York] epitomized the gallantry of American fighting men and their sacrifices on behalf of freedom."

BOOKS

Sergeant York And His People by Sam K. Cowan

INTERNET

Here's a clip of the movie *Sergeant York*, which shows the exchange with Major Buxton: youtube.com/watch?v=eyJTfuau3hw

The Doughboy Center collects the letters, diaries and other stories of the American Expeditionary Forces who served in WWI. You can read some of them here:
www.worldwar1.com/dbc/biograph.htm

MOVIES

Sergeant York (1941)

The Battle of Cantigny

May 28, 1918

CANTIGNY: WHERE THE AMERICANS WON THEIR FIRST LAURELS

EARLY on the morning of May 28, 1918, three battalions of our 28th Infantry, First Division—the first to land in France—swept out of their trenches and, aided by French tanks and Franco-American artillery, captured the strongly held village of Cantigny. A tense moment in that fight has been vividly caught by Mr. Schoonover, who worked from official photographs and descriptions by participants.

1918 artist's depiction of the Battle of Cantigny

YOU WON'T FIND the Battle of Cantigny listed among the world's most important battles.

But this battle was really important for the United States, mostly because it had so many firsts.

- The first time American Expeditionary Forces (AEF) led a division-sized offensive in WWI.
- The first WWI battle led by an American commander.
- The first American victory of WWI.
- The soldiers were part of the US Army's First Infantry Division.

American soldiers attack at Cantigny, France. Photo courtesy of First Division Museum Research Center

On April 22nd, 1918, the US First Division moved to the Western Front near the Somme River in France. The Americans quickly realized their defensive positions didn't have enough protection to keep them safe from the constant German artillery firings. Instead of panicking, the Americans dug in. By mid-May they had completed a stronger defensive line. On May 27, the American regiment was moved into front line positions and ordered to take the town of Cantigny, which was controlled by the Germans.

Cantigny itself wasn't an important military asset. But the Germans were planning to use the village as a launching pad to reach Paris, which was only about fifty miles away. If the Germans took control of Paris, the war's momentum would shift dramatically toward the Central Powers.

In the meantime, German gunners used Cantigny's high ground both as a lookout over the region and a good place from which they could fire on the Allies.

As the Americans came toward Cantigny, German General Erich Ludendorff ordered his army to utterly destroy them. Ludendorff wanted to prove to the world that these new soldiers would be no help to the Allies.

The Americans were new to warfare. Before this battle, the Allies only used the American Expeditionary Forces (AEF) as reserve units that would help out other armies. Nicknamed "Doughboys" by the Allies, the AEF had never fought under an American commander. But at Cantigny, they would follow orders from US General John "Blackjack" Pershing.

US General John "Blackjack" Pershing in France 1918

The untested Americans were going up against three experienced German regiments. To make up for being outnumbered by veteran fighters, the Doughboys were ordered to strike fast and use massive firepower. To that end, the French supplied them with tanks and flamethrowers.

WWI US soldier, otherwise known as a Doughboy.

The battle started out badly for the Doughboys.

On the night of May 24, an American officer got lost in no man's land. The officer was carrying maps of the Americans' positions, and Germans captured him—and the maps. Four days later, when the Allies opened fire in the pre-dawn hours, the Germans immediately countered with their own artillery barrage.

The barbed wire fields of WWI

Around 6:30 a.m., the Doughboys moved out of their trenches under protection of a rolling barrage—artillery guns firing ahead of the infantry to wipe out forward enemy positions.

Most of the American soldiers headed west toward Cantigny. A smaller force moved in from the south. The French also came forward with rifle and machine-gun fire, including a platoon of flamethrowers and a dozen M-16 French tanks that helped destroy about twenty German machine-gun positions.

French flamethrowers (left) attack German trenches

As the Americans moved forward, the Germans' murderous shots cut men to the ground. But some of them managed to reach the town of Cantigny, including Sergeant Boleslaw Suchocki of the 1st Division.

"There remained nothing but ruins," he later recalled. "We passed on through to the other side of the village. Here we encountered barbed wire entanglements but it was our good fortune to get through these without any mishap. But once across I notice that the boys are falling down fast. A shell burst about ten yards in front of me and the dirt from the explosion knocked me flat on my back. I got up again but could not see further than one hundred feet...

"The German artillery was in action all the time... I stopped at a strong point and asked the boy in the trench if there was room for me to get in. 'Don't ask for room, but get in before you get your [! %&] shot off,' a doughboy said...."

French flamethrowers clear a bunker,
photo courtesy of the First Division Museum

Within several days, the Americans took control of Cantigny and captured a large number of prisoners.

The Germans counterattacked.

Now the American objective was to hold the town, keeping

it from falling back into German control. The Doughboys dug into defensive positions. Meanwhile, however, the Germans were already advancing on the nearby town of Château-Thierry. This threat prompted the French to withdraw from Cantigny; they needed to protect their troops at Château-Thierry.

The Americans were on their own.

No more heavy artillery.

No more tanks.

And no more soldiers with experience on the Western Front.

Inside the defensive positions, the Doughboys positioned their automatic rifles and machine guns, laid out more barbed wire, and kept up constant patrols.

All the while, the Germans continued to fire on them—and added some chemical gas attacks, too. The poisonous clouds wafted through the air, burning through Cantigny.

Machine gunners wearing gas masks

Battered and exhausted, the Americans kept repelling the German infantry assaults. The bombardments grew so violent that soldiers were buried under rubble. As the American holdout continued, the Germans stepped up their attacks even further, desperately trying to prove that these American soldiers couldn't help the Allies win the war.

Then a sudden change.

A German communications breakdown forced their artillery to fire at the wrong time—in fact, the Germans were firing right as their ground troops attacked, killing their own men! Despite the uncoordinated attack, the Germans still managed to carve their way toward the Americans.

But the Doughboys refused to quit. Holding their ground, firing their rifles and machine guns, the Americans proved steadfast. By May 30, nearly 200 American soldiers were dead, another 852 were wounded, and sixteen were missing or captured. And yet, the Americans managed to destroy German positions and finally force the enemy to pull out of Cantigny.

The Battle of Cantigny was over—and the Doughboys had won!

News of this hard-won American victory rippled through the Allied forces. And it shifted some attitudes. For instance, before Cantigny, French enlisted soldiers thought so little of the Americans that they didn't even bother saluting the US officers.

After Cantigny, they did.

WHO FOUGHT?

Brigadier General Theodore Roosevelt, Jr.
shortly after his landing at Normandy, France, 1944

Teddy Roosevelt, Jr. heard a lot of history lessons from his father—US President Theodore Roosevelt.

President Roosevelt was a famous military veteran who had fought bravely in the Spanish American War of 1898.

"During every battle [lesson] we would stop and father would draw out the full plan in the dust in the gutter with the tip of his umbrella," Teddy recalled. "Long before [WWI] had broken over the world father would discuss with us military training and the necessity for every man being able to take his part."

As soon as the United States declared war on Germany, the former president sent a note to Major General John "Blackjack" Pershing. President Roosevelt asked if his three sons could join the American Expeditionary Forces that were headed to Europe.

Teddy Jr. was among the first Americans to fight on the Western Front. As a battalion commander, he led his men into combat, facing down artillery fire and chemical gas, including at the battle of Cantigny. Teddy Jr. also purchased combat boots for all his men—with his own money. For his actions on the WWI battlefield, he was awarded the Distinguished Service Cross.

Unfortunately, his brother Quentin, a pilot, was killed in battle.

When WWII broke out more than twenty years later, Teddy Jr. returned to active duty in Europe. In 1941, he was given command of a regiment in the 1st Infantry Division—the same unit that he'd fought with during WWI.

In 1944, the WWII Allies landed at Normandy Beach. It was a deadly mission known as D-Day and Teddy Jr. was the only general among the first wave of troops that reached the French beaches controlled by Nazi soldiers. He was 56 years old, the oldest soldier in the entire invasion.

Another Roosevelt also landed at Normandy that day— Teddy Jr.'s son, Captain Quentin Roosevelt II, named after his pilot-uncle who died in WWI.

For his heroic actions on D-Day, Teddy Jr. earned a nomination for the Distinguished Service Cross—the same medal he'd won during WWI. But about a month later, Teddy Jr. suffered a heart attack and died.

He was buried at an American cemetery in Normandy, France dedicated to Americans killed in the D-Day invasion.

Although Teddy's brother Quentin was buried in a different French cemetery, his remains were later moved to the same Normandy cemetery, allowing the Roosevelt soldier-brothers to rest in peace beside each other.

BOOKS

The Battle of Cantigny (Cornerstones of Freedom Second Series) by
   Tom McGowen

INTERNET

The 1st Infantry Division of the US Army has taken part in nearly
   every American War since WWI. The First is the longest
   continuously-serving division of the US Army. The division's
   nickname is "The Big Red One" because of their red shoulder
   patch that displays a large numeral 1. Sergeant Suchocki's
   thrilling account of being at the Battle of Cantigny comes
   courtesy of the First Division Museum in Illinois. The museum
   offers many good resources. They also have a tank park! Check
   out all the different models here:
   www.fdmuseum.org/exhibits/c/tank-park

MOVIES

This 1918 newsreel shows authentic battlefield footage from
   Cantigny. Watch the artillery bombardments as they happen!
   archive.org/details/27034aCantignyMosVwr

Harlem Hellfighters

The 369th in action during WWI. Painting by H. Charles McBarron.

WHEN WWI BEGAN, the American military had some bad rules which were common throughout America.

These rules fell under the term "segregation."

Segregation ("seg-reh-gai-shun") divides people into different social groups based only on some unimportant criteria, such as the color of a person's skin. Although the American Civil had helped to end slavery, black Americans were still segregated from white Americans—even in the US military.

The Army had segregated black and white regiments. Black

regiments were commanded by white officers. These so-called "colored" soldiers served in both infantry and cavalry, and fought out West in the "Indian Wars." Black soldiers also fought in the Spanish-American War of 1898. Nicknamed Buffalo Soldiers, they proved to be legendary fighters (and their commander in that war was none other than Teddy Roosevelt, before he became president of the United States.)

Lieutenant Wesley Herbert Jamison of 351st Machine Gun Battalion, 92nd Infantry Division

By 1917, the United States Army had four all-black regiments: the 9th and 10th Cavalry and the 24th and 25th Infantry. After Congress declared war on Germany, about 400,000 black American men were drafted into the Army, Navy, and Coast Guard. They were not allowed to serve in the Marine Corps.

Remember Sargeant York's draft card? Here it is again. This time, look at the bottom left corner. It reads, "If person is of African descent, tear off corner."

That torn-off corner told the draft board whether the recruit was black. Skin color would determine where a soldier served. During WWI, most black recruits were assigned manual labor jobs—cooks, janitors, construction workers—but about 40,000 served in combat overseas.

WWI's most famous black combat unit was the 369th Infantry. These men proved courage doesn't have a skin color.

The 369th Infantry during WWI

While other black units in Europe were given support duties, such as loading trains and ships, the 369th went straight into combat. They were welcomed heartily by the French Army, which by 1917 was running low on men and morale. The French even equipped these black Americans with French firearms and helmets.

The 369th kept their own motto: "Don't tread on me."

Black troops of the 351st Field Artillery, 92nd Division, 1918

The 369th soldiers, many of whom grew up in the New York City neighborhood of Harlem, served in WWI trenches longer than any other American unit—a total of 191 days. They also never lost a prisoner and never gave up one foot of captured ground. Their courage impressed even the Germans, who nicknamed them the "Harlem Hellfighters."

Here's an example of how the 369th earned that nickname. The Germans liked to sneak into Allied territory on night raids, killing unsuspecting soldiers and stealing food and other supplies. During one such raid, US Corporal Henry Johnson of the 369th fought off the entire German raiding party—at one point using only his knife for hand-to-hand combat. Johnson suffered 21 wounds but refused to give up.

For his actions, France awarded Johnson its Croix de Guerre medal, given to individuals who distinguish themselves by acts of heroism in combat with enemy forces. Johnson was among the first Americans—black or white—to receive this high military honor.

Nearly 80 years later, after Johnson passed away, America awarded him the Purple Heart and Distinguished Service Cross. His heroics were described in detail at the awards ceremony.

"Private Johnson distinguished himself by extraordinary heroism while engaged in military operations involving conflict with an opposing foreign force. While on a double sentry night duty, Private Johnson and a fellow soldier were attacked by a raiding party of Germans numbering almost twenty, wounding both. When the Germans were within fighting distance, he opened fire, shooting one of them and seriously wounding two more. The Germans continued to advance, and as they were about to be captured Private Johnson drew his bolo knife from his belt and attacked the Germans in a hand-to-hand encounter. Even though having sustained three grenade and shotgun wounds from the start, Private Johnson went to the rescue of his fellow soldier who was being taken prisoner by the enemy. He kept on fighting until the Germans were chased away. Private Johnson's personal courage and total disregard for his own life reflect great credit upon himself, the 369th United States Infantry Regiment, the United States Army, and the United States of America."

The other Harlem Hellfighters showed similar courage under fire—so much so that at the end of WWI, France awarded it Croix de Guerre to the entire unit!

Back home in America, people learned of the 369th's bravery and hard fighting, and it helped change opinions about segregation, especially in the military.

By the end of WWI, black soldiers were serving in the cavalry, infantry, signal corps, medical, engineer, and artillery units. Other men worked as chaplains, land surveyors, truck drivers, chemists, and intelligence officers.

Unidentified American soldier in uniform
with First Army shoulder insignia and campaign hat

Perhaps one of the best examples of how public perception had changed was the parade given for the 369th.

In 1917, when the 369th first left New York to serve abroad, the men weren't given a farewell parade, even though all other departing military units had one. But in 1919, when the Harlem Hellfighters returned home, massive crowds filled New York City's Fifth Avenue. People of all races wanted to celebrate these soldiers and thank them for their bravery on the battlefield. In Harlem, schoolchildren were dismissed for the day to welcome home such highly decorated hometown heroes.

Thousands of spectators in New York City, 1919,
welcoming home the 369th Infantry

However, full equality for black Americans was still a long time coming—and another parade proved that point.

After the Allies won WWI, New York City threw a Victory Parade to celebrate the American servicemen who helped bring victory over the Central Powers.

But the Harlem Hellfighters were not part of that parade.

BOOKS

The Harlem Hellfighters by Max Brooks and Caanan White

The Harlem Hellfighters: When Pride Met Courage by Walter Dean Myers and Bill Miles

Harlem Hellfighters by Shannon Baker Moore

Walter Tull. WW1 Black Officer & Footballer by Alex Devaney

INTERNET

History.com has a video on the Harlem Hellfighters:
www.history.com/topics/world-war-i/the-harlem-hellfighters-video

Belleau Wood

June 1–26, 1918

American soldiers fire a 37 mm French artillery gun,
maximum range more than one mile.

AFTER THE BATTLE of Cantigny, the American Expedition Forces were given another objective. The Germans held a forested area in France known as Belleau Wood (pronounced "bell-oh wood"). The United States Marines, backed by army artillery, were ordered to drive them out and take control of the woods.

The mission was a tall order.

After Russia dropped out of the war in 1917, Germany moved its forces off the Eastern Front. This shift added about

500,000 soldiers—fifty divisions—to the Western Front. By March 1918, the Germans were creeping ever closer toward their goal of Paris. Belleau Wood, for example, was only about fifty miles from the French capital.

On June 1, 1918, the US Marines arrived at Belleau Wood and dug in, taking defensive positions for the upcoming battle. Some of these positions were only deep enough for a man to lie on his stomach and fire his rifle. Meanwhile, other Allied forces, including the French army, gathered nearby for support.

The Germans marched forward. The Allies held their fire. The Germans came closer. And closer.

When the Germans were about 100 yards out, the Allies opened fire.

WWI American troops

But the Americans on the front line didn't have enough support from the Allied assault companies. Soon the Germans were overpowering them. As the firefight continued, French troops started retreating—and they hollered at the Marines to get out of there, too.

"Retreat?" cried Marine Captain Lloyd Williams. "Hell, we just got here!"

On June 4, the Marines still held the line, and the French decided to send some troops back to support them. Two days later, in the early morning hours, the Marines heard Allied artillery opening fire on the German positions. Unfortunately, this pre-bombardment only alerted the Germans to an oncoming attack.

The Marines waited inside their trenches, listening for their commander's whistle. When it blew, they went over the top.

Two US soldiers race toward a bunker,
stepping over the remains of two German soldiers.

Bayonets ready, the Marines raced forward, coming within 50 yards of the enemy when the Germans opened the gates of hell. Machine guns, rifles, artillery, snipers—all of that fire-power slaughtering about one thousand Marines.

At that point, this Marine assault should've turned into a total massacre. But one young captain, George Hamilton, took control. Grabbing some of his men, Captain Hamilton made a break for the nearby woods. The rest of the Marine company kept fighting forward, despite the heavy fire against them.

LA BRIGADE MARINE AMERICAINE AU BOIS DE BELLEAU

Draft de GEORGES SCOTT.

Artist's rendering of the American Marines at Belleau Wood

Hamilton and his men discovered an area that allowed them to see the Germans firing from machine-gun nests. The Marines charged them with bayonets, slashing into the enemy and keeping up the fight until they overtook the top of a hill. The German lines broke and ran.

The Marines had gained a foothold in Belleau Wood.

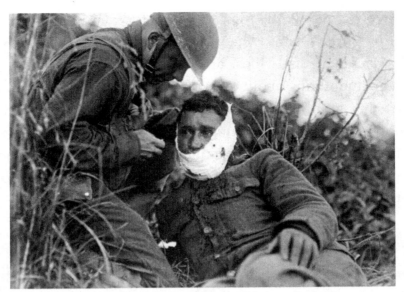

Wounded American soldier receiving medical aid

Several days later, on June 10, the Marines moved north through the woods with a new objective, the town of Bouresches ("Boor-eh-shez"). But as the Marines came in for the attack, the Germans hit them with artillery and machine-gun fire. The Marines kept pushing forward, even as men dropped dead like cut trees. Although 150 men started this trek to Bouresches, when the Marines reached the town, only about twenty were still alive.

The Marines began searching for the enemy inside the town. Germans hid inside attics and cellars, behind doors and walls, all of them waiting for their chance to kill the Americans.

But the Marines were not giving up. They had fought their way through countless obstacles, and now their marksmanship training would prove its worth. They soon cleared the town of all enemy combatants, and held their ground until reinforcements arrived.

In the early morning hours of June 10, the Marines were

ordered to leave Bouresches. Firing their rifles and throwing hand grenades, they covered a quarter-mile of open ground and eliminated more German machine-gun nests. By 6:00 a.m., they were back in Belleau Wood.

American soldier killed in action among barbed wire

This single battalion of Marines couldn't clear the entire woods of Germans and a second Marine battalion was sent in. By June 12, the Germans had fired more than 5,000 artillery rounds at the Americans. Just as dangerous, the Marines were now facing food and water shortages, along with a severe lack of sleep. The Germans then added deadly chemical gas attacks, forcing the Americans to fight wearing heavy masks. When reinforcements showed up, they soon found themselves in hand-to-hand combat, fighting Germans using their fists and bayonets. It was no wonder that the German soldiers named these ferocious Marines "teufel hunden"—which translates as "devil dogs" or "dogs from Hell."

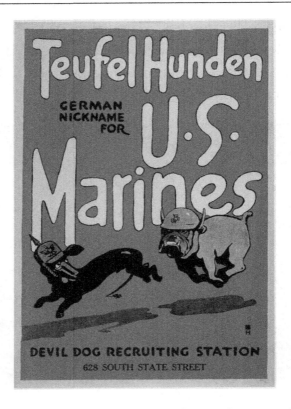

The Germans finally retreated from the woods—only to regroup and counterattack six times.

Each time, the Marines held.

The Germans realized they couldn't win this battle. They retreated.

"The deadliest weapon in the world," General Blackjack Pershing once said, "is a Marine and his rifle."

The Battle of Belleau Wood was over—and the Marines had won.

Unfortunately, this battle inflicted the highest number of Marine casualties in history up to that time—thirty-one officers, more than 1,000 soldiers, with nearly 10,000 men wounded.

Although German casualties were never reported, about 1,600 soldiers were taken as prisoners.

The Battle of Belleau Wood marked another turning point, for the Americans in particular and for the war in general. The Marines won the battle, and stopped the German advance on Paris.

The American Expeditionary Forces were shifting the war's momentum toward Allies. Victory was within reach. The Central Powers were about to fall.

WHO FOUGHT?

CAPT. LLOYD W. WILLIAMS, U.S.M.C.

Lloyd Williams was born in Berryville, Virginia on June 5, 1887.

Shortly after graduation from Virginia Polytechnic Institute, where he was a member of the Corps of Cadets, Williams was commissioned as a second lieutenant in the United States Marine Corps. Typical of a career officer, Williams spent time at the Marine Corps Headquarters in Washington, D.C. then served tours of duty aboard American battleships everywhere from Cuba to the Philippines.

When the United States entered World War I, Williams and

the 5th Marine Division sailed for France. Williams was assigned command of the division's 51st Company in the 2nd Battalion.

On June 11, 1918, he led the assault at Belleau Wood. Williams was wounded on the battlefield But when the medics tried to help him, Williams said, "Don't bother with me. Take care of my good men."

Although Williams was evacuated to a medical station—that was getting shelled by the Germans—he died of his wounds. After his death, he was awarded the Distinguished Service Cross. His body was buried in "Flanders Field," a military cemetery near Belleau Wood.

In 1921, the Williams' family had his remains returned to the United States and buried in his hometown of Berryville, Virginia. Such was Williams' fame that national newspapers reported on his casket as it was loaded onto a Navy ship in France, and upon its arrival in New York. A military honor guard escorted the coffin to Virginia. On July 21, 1921, more than 1,000 people filled the streets of Berryville to honor Major Williams.

In the history of the US Marine Corps, the 2nd Battalion, 5th Regiment is the most highly decorated battalion. Today, its motto remains "Retreat, hell!"—a shortened version of what Captain Williams yelled to the French who were retreating from the battlefield at Belleau Wood.

BOOKS

A Regiment Like No Other: The 6th Marine Regiment at Belleau Wood by U.S. Army Command and General Staff College

Château Thierry & Belleau Wood 1918: America's baptism of fire on the Marne by Dávid Bonk and Peter Dennis

US Marine vs German Soldier: Belleau Wood 1918 by Gregg Adams and Steve Noon

INTERNET

Here is a good video presentation of the Battle of Belleau Wood: youtube.com/watch?v=ZsgwHxboHFk

MOVIES

Devil Dogs: Hero Marines of WWI

World War I: American Legacy

AFTERMATH

Signing WWI's peace treaty in the Palace of Versailles Hall of Mirrors.

By late 1918, Germany was facing a grim reality.

Several members of the Central Powers—including Bulgaria, Austria-Hungary, and the Ottoman Empire—had signed armistices with the Allies. An armistice ("arm-ah-stiss) isn't necessarily the end of a war, but it is a formal agreement among warring parties to stop fighting and negotiate some kind of lasting peace.

The Central Powers were also broke. No money remained to sustain this war, and inside Germany and Austria-Hungary, people were starving.

On November 11, 1918, Germany signed an armistice.

World War I was officially over.

The Germans had hoped to prove the American forces made no difference in this war. But the truth was now clear. With the United States joining the war, the Allies had received a fresh supply of men and material, an influx that increased numbers on the battlefield and boosted morale among the other troops. Also, in battles such as Cantigny and Belleau Wood, the Americans showed they would fight to the death— no retreating.

Worn-out and bankrupt, the Central Powers could no longer compete.

American Expeditionary Forces march through London, England

But the Allied victory came with huge costs. Entire European towns had been destroyed, leaving nothing but rubble. About ten million young men were dead—killed by enemy fire, mortally wounded, or their lives taken by a deadly disease. Adding to this "lost generation" was another ten million civilians who suffered and died in the war.

WWI offered many lessons. Devastating conflicts sometimes

arise from a political lust for power. Dangerous alliances can wreck international peace. And national pride can take a terrible turn when one country decides it should rule over other countries. We've also seen how poor military leadership, modern weaponry, and devastating tactics such as trench warfare can wipe out millions of people.

The Treaty of Versailles was signed on June 28, 1919— exactly five years after the assassination of Archduke Franz Ferdinand, the event that triggered World War I. The Allies, as the war's victors, divided up parts of Europe previously held by the Central Powers. Six new countries sprang up: Poland, Czechoslovakia, Yugoslavia, and Estonia, while Austria-Hungary became Austria *and* Hungary.

Here's a map that shows these changes. Look at the new borders drawn over the old countries.

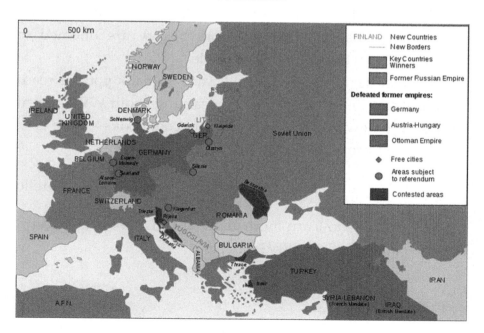

The Treaty of Versailles was supposed to bring world peace. But its requirements were really harsh. For instance, the Treaty gave nothing to the Central Powers. In fact, Germany was ordered to repay all the debt racked up by the Allies because of this war. That requirement was a tall order for a country that was already broke.

Some Allied leaders even insisted this treaty would only make things worse—for Germany and everyone else.

These skeptics were right.

The treaty's demands only further angered the German people, and their national rage helped fuel the rise of a ruthless political leader. His name was Adolf Hitler. During the 1930s, Hitler and his National Socialist political party—otherwise known as the Nazis—seized control of Germany. By 1939, Hitler had launched World War II.

That second world war is another amazing part of history, and you should learn about it. In fact, you should never stop learning about history in general. As Winston Churchill once said, "Those that fail to learn from history are doomed to repeat it." Don't be one of those people!

As we close out this journey, I want to thank you for reading this book. I'm proud of you for learning about WWI's great battles. And now you can go read more cool books—there is so much more history for you to discover!

—Joe Giorello

P.S. If you'd like to learn more about WWII, here is an excerpt of *Great Battles for Boys: WW2 Europe.*

First Chapters of *WW2 Europe*

PRELUDE TO WAR

Adolf Hitler, 1937

"WARS AND RUMORS of wars."

People have said that phrase for thousands of years. But what does it mean?

It means that wars don't just fall from the sky. Major conflicts erupt from a sequence of events, and people talk about those events—the "rumors of war"—long before actual war breaks out.

The "rumors of war" that led up to WWII actually began with World War I.

Known as "The Great War," not because it was so good but because millions of men lost their lives, WWI lasted four years, from 1914–1918. Some 37 million people were killed or wounded. The reason behind WWI were complicated, but everyone agrees on these facts: Germany started WWI, the war nearly wrecked Europe, and four years later, Germany lost WWI. And because the war caused so much pain and suffering, France wanted to punish Germany for starting the war. So France added some harsh terms to the peace agreement, called The Treaty of Versailles. This treaty ordered Germany to pay back a huge amount of money to the countries that won WWI, including France.

By itself, that demand wasn't unusual—the winning side usually makes the losers pay back some of the debt. The process is called "reparations."

But Germany was broke after WWI. The German people were starving. They had no jobs and the Germany currency—its money—was almost worthless because the government had gone bankrupt.

Several countries, including the United States, thought France should lighten up on the terms forced upon Germany. After all, how could Germany come up with extra money for reparations when it couldn't even feed its own people? American and British leaders also worried that the peace treaty's harsh terms would create even more bitterness among the German people who would resent the winning countries, planting seeds for more fighting.

France refused to listen.

So the Treaty of Versailles was ratified, or made official. Germany was forced to obey all of the treaty's demands, including surrendering some of its land.

Sadly, just as predicted, the treaty made matters worse.

Germany never climbed out of its economic depression. And the German people grew only more furious because of the treaty's terms. Now they hated France even more than before WWI had broken out. And they also hated the other winning countries, such as Great Britain and the United States.

This national bitterness increased throughout the 1920s and 1930s, and into that terrible resentment walked a ruthless political leader. His name was Adolf Hitler.

Crowds cheer Hitler who stands in the window after being inaugurated as Germany's chancellor on January 30, 1933.

During World War I, Hitler served in the German army. He was even awarded several medals for bravery. After the Great War, Hitler rose through the German government, gathering power among former soldiers like himself who hated the Treaty of Versailles.

But not all Germans supported Hitler. Some politicians threw him in jail.

While he was locked up, Hitler wrote his autobiography, *Mein Kampf* (pronounced "mine kahm-ff"). The title translates to "My Struggle." Hitler used this book to tell the German people that they were extra special, that they belonged to something called the Aryan race. Aryans had light-colored skin, blonde hair, and blue eyes. Hitler said Aryans were superior to all other races and it was their "historic destiny" to rule over the entire world.

Hitler also said Germany needed to get rid of its less-worthy people, especially Jews, Gypsies, and Polish people. These people, he insisted, were holding back Germany and keeping it from its rightful place of world domination. If Germany would "cleanse" itself of these lesser people, there would be no more hunger, no lack of jobs, and no more submitting to "inferior" countries such as France.

The German people, tired of being poor and hungry, liked what Hitler was saying. They wanted better lives, and Hitler was offering them hope and change. They believed Hitler would lead their country to freedom and prosperity.

German children were taught to worship Hitler.
Here, schoolchildren give the Nazi salute. 1934

Prison had only made Hitler more popular, and when he was released, he had even greater political power. Also, Hitler learned to became a great public speaker by watching the speeches of Italian dictator Benito Mussolini, leader of Italy's Fascist Party. Mussolini drew huge crowds with his passionate speaking style. Both Hitler and Mussolini held many of the same ideas, especially about the government having total control over its people. Both men wanted to be dictators. Soon, Hitler and Mussolini formed an alliance.

Hitler's political party was the National Socialists or "Nazis." The socialists started winning more and more political offices. Hitler created military-style organizations within the government. He chose leaders who believed Germany should rule over the entire world and "cleanse" itself of all non-Aryan races, particularly the Jewish people.

Hitler and his Nazi party soon won control of the entire German government.

In his speeches, Hitler kept painting a picture of the new Germany. He called it The Third Reich and said it would "last for a thousand years." But to build The Third Reich, Hitler needed to reclaim the land that was taken away from Germany through World War I's Treaty of Versailles.

Hitler was particularly focused on land given to create the country of Poland, and his complaints were echoed in grumblings by the German people—Poland had "stolen" German land. These complaints were among those "rumors of war," and Poland became one of Hitler's first targets.

When he struck, all "rumors of war" vanished into an explosive assault.

World War II had begun.

THE INVASION OF POLAND

September 1, 1939

German soldiers marching into Poland, 1939

HAVE YOU EVER watched a little kid playing with a toy he really likes? If you suddenly take away that toy, the kid gets upset—naturally.

But what if you not only take away that toy, but you give it to somebody the kid doesn't like? In fact, you give his toy to his enemy.

Now that kid's *really* upset.

That's similar to how Germans felt when the Treaty of Versailles forced them to surrender some of their land. And not only was their land taken, but it was also given to Poland. The Germans didn't like the Poles.

Look at the map below of 1939 Europe. Germany sits in the middle, marked in dark gray, and to the immediate right is the land taken from Germany to create Poland, which was now Germany's neighbor. Furthest to the right is the country of Russia, then known as the Soviet Union.

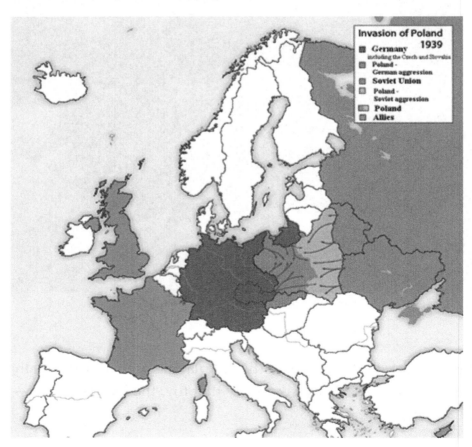

Hitler cleverly played on the emotions of the German people. He said they should be mad—how dare France and other countries take away that land! How dare the Polish people live on it! That land belonged to Germany.

But as Hitler was saying all these things to his own people, he was also fooling the rest of the world. For instance he signed a "non-aggression pact" with Poland. With this agreement, Hitler promised not to attack Poland.

Hitler's actions confused some of the German people. How was Germany going to reclaim that land if Hitler was promising not to invade Poland?

What they didn't know, this double-talk was all part of Hitler's plan.

In the summer of 1939, several years after signing the non-aggression pact, Hitler and other German politicians started accusing Poland of all kinds of terrible things. German newspapers reported these accusations like they were facts. That's called propaganda—when false information is reported as truth to further a political cause.

As part of their propaganda, Hitler and his henchmen accused the Polish government of killing innocent Germans. To "prove" it, on August 31, 1939, some German spies dressed up like native Poles seized a Polish radio station. The spies broadcast an ugly report about Germans—all spoken in Polish—then dragged some dead bodies to the radio station to make it look like Poland really was killing innocent Germans.

That same night, the spies carried out even more "crimes" against Germans, including setting a house on fire.

These events, naturally, infuriated the German people.

Of course, we now know all these crimes were based on lies and deception. But at the time, they were part of "Operation

Himmler," Hitler's plan to invade Poland. Yes, Hitler had signed that non-aggression pact but now he could claim that Poland broke the pact by killing "innocent Germans." See how clever Hitler was?

Hitler didn't care that the accusations and crimes were based on lies. "The victor," he said, "will not be asked whether he told the truth."

The day after those radio station "murders," Operation Himmler launched its next phase.

The German air force, called the Luftwaffe, started bombing Polish cities, bridges, roads, airfields, and communication centers.

Thousands of innocent Poles were killed.

Right after the Luftwaffe attacked, German tanks smashed through Polish villages, towns, and markets. These steel machines were followed by the German infantry, shooting and killing thousands more Polish people.

The date was September 1, 1939.

World War II had officially begun.

Two days later, on September 3, the countries of Great Britain, France, Australia, and New Zealand declared war on Germany.

Unfortunately, for Poland, the declaration was too late to save the country.

Polish artillery after German air attack

Even though Hitler had signed that non-aggression pact, Poland never really trusted him. Polish leaders suspected Hitler might try something sneaky, but nobody ever expected such a sudden and ruthless attack on innocent Polish civilians—a civilian is someone who is not a soldiers or part of the military.

Germany was launching a new style of warfare, called *blitzkrieg*, or "lightning war." It worked like a sucker punch, coming out of nowhere and striking so quickly and so powerfully that the victim couldn't recover in time to defend himself.

As the Luftwaffe bombers and fighter planes smothered the skies over Poland, massive concentrations of armored and motorized infantry stormed the streets, shattering all defenses.

Blitzkrieg not only surprised an enemy, it also created so much chaos that it was difficult for the opposing army to organize its troops. Making matter worse in this invasion, the Polish army was still using weapons and tactics from WWI.

Polish cavalry soldiers spurred horses and swung swords against German armored tanks and machine guns. Polish biplanes made of canvas and wood struggled to outshoot modern German fighter planes, such as the Messerschmitt Bf 109.

Polish cavalry soldier, 1938

And yet, even with all that against them, Polish soldiers fought courageously. Three squadrons of Polish cavalry soldiers attacked Germany's 8th Army, forcing them to retreat. The horseback-riding soldiers pursued the Germans through heavy machine-gun fire. The Polish cavalry also managed to capture one of Germany's divisional headquarters, taking as prisoners one general and about 100 soldiers. Another Polish cavalry squadron defeated a German cavalry unit—it was one of the last all-cavalry battles in military history.

But the blitzkrieg had confused Poland's military commanders. They sent troops to the wrong defensive positions. German ground forces broke right through the lines.

Lone horse, stranded in a Polish battlefield, September 1939

Then, on September 17, the Soviet Union invaded Poland, too.

Look at that map of Europe (above) again. The arrows to the far right show how Soviet forces invaded Poland from the east. Turns out, Hitler had yet another mean trick up his sleeve. He had secretly signed an agreement with Josef Stalin, the leader of the Soviet Union. These two dictators agreed to invade Poland and divide the country between themselves. They also agreed not to attack each other.

As the Soviets invaded, Poland's commander-in-chief fled into Romania.

But the attack still wasn't over.

Unleashing its first large-scale aerial bombardment on a major city, the Luftwaffe struck Bzura. For ten days, out-manned Polish forces tried to fight back, but Germany's 8th Army was advancing. Every Polish counterattack failed.

The Luftwaffe also bombed all the bridges across the Bzura River, trapping Polish forces and leaving them vulnerable to German Stuka planes that strafed them with fire from above.

Finally, as their anti-aircraft ammunition ran out, the Polish forces retreated into some nearby forests. The Germans pursued them while planes such as the Heinkel He 111s and Dornier Do 17s dropped still more incendiaries, smoking the Poles out of the canopy of trees.

In this one campaign, Germany dropped hundreds of tons—possibly even thousands—of bombs.

Young survivor of the Luftwaffe bombing of Warsaw, Poland, 1939

Other Polish cities also burned to the ground, and the Luftwaffe continued to target civilians, even refugees fleeing the fighting. These mass killings were all part of the blitzkrieg's tactics.

In addition to these terrifying attacks, German secret police were murdering thousands more people, sending terror through the entire population.

In this one September campaign alone, about 150,000 Polish citizens died.

Finally, realizing there was no way to win, Poland surrendered to Germany on September 27, 1939.

About 300,000 Polish soldiers were captured, becoming prisoners of war. And, just as they planned, Hitler and Stalin divided Poland. Germany took over the western half of the country, the Soviet Union grabbed the east.

Both dictators then turned their sights on the rest of Europe. Hitler and Stalin wanted more land.

And they would kill to get it.

WHO FOUGHT?

Heinrich Himmler

The secret plan to launch WWII was code named Operation Himmler, after Heinrich Himmler, one of Hitler's leading henchmen.

Himmler joined the Nazi party in 1923 and helped Hitler rise to power. Early on, Himmler coordinated Nazi propaganda. When Hitler needed an excuse to invade Poland, he turned to Himmler, who came up with those false stories for the German newspapers and coordinated the fake murders at the Polish radio station.

Himmler considered non-Aryans "inferior" races. He believed in the "extermination" of these people so that Germany could become "pure." Himmler set up and controlled concentration camps where he ordered the mass murder of millions of innocent people—specifically Jews.

As a child growing up in Germany, Himmler belonged to the Catholic church. But his devotion to Hitler persuaded him to abandon his Christian religion. Himmler then became obsessed with the occult or "dark magic." For instance, as part of his propaganda duties, Himmler designed symbols used by the Nazis to identify themselves. One Nazi group, handpicked by Hitler, was called the "SS." It stood for *The Schutzstaffel* or "Protection Squadron." The SS soldiers were cruel beyond imagination. Himmler created the SS symbol based on ancient Greek mythology—two lightning bolts, side by side. The SS wore this symbol on their uniforms, striking fear in their victims.

BOOKS

Blitzkrieg: The German Invasion of Poland and France 1939 to 1940 by Phil Yates

Hitler Youth: Growing Up in Hitler's Shadow by Susan Campbell Bartoletti

World War II: An Interactive History Adventure by Elizabeth Raum

DK Eyewitness Books: World War II by Simon Adams

INTERNET

Read the diary of a Polish doctor who witnessed the Nazi takeover of his country: www.eyewitnesstohistory.com/poland.htm

This short documentary on YouTube shows historical footage of the invasion: youtube.com/watch?v=uNOqSSP1o94

Watch the Blitzkrieg in action: youtube.com/watch?v=vgCWMZaKKUw

MOVIES

The Waffen SS: Hitler's Elite Fighting Force

WORLD WAR ONE ALLIES

(in alphabetical order)

Australia

Canada

France

Great Britain (England, Scotland, Wales, Ireland)

Greece

Italy

Montenegro

Portugal

Romania

Russia

Serbia

United States

WORLD WAR ONE CENTRAL POWERS

Austria-Hungary

Germany

The Ottoman Empire

Bulgaria

GLOSSARY

Here are some military terms and definitions.

ambush: a surprise attack, usually from a concealed position.

armistice: a formal agreement between warring parties to stop fighting.

artillery: Cannons, howitzers, large-caliber guns

blockade: sealing off a place to prevent goods and/or people from entering or leaving.

creeping barrage: Firing units remain in the same relative position while another force advances under its cover.

delaying action: a defensive force tries to delay the advance of a superior enemy force by withdrawing and at the same time inflicting the maximum destruction possible, all while not engaging in decisive combat.

dreadnought: a type of battleship introduced in 1906 that was larger and faster than its predecessors and was armed with heavy-caliber guns in the turrets. These ships were named after the British battleship *Dreadnought*, the first of its kind.

fleet: a large formation of warships or planes under one command.

foxhole: a small pit dug in a battle area to protect a small number of soldiers.

garrison: A military post, usually permanent. The word can also be used for the troops stationed at a military post.

magazine: Storehouse for ammunition. Also, a small metal container that holds bullets for certain types of automatic weapons.

mobilization: how an army moves its supplies, especially food and ammunition, from one place to another. Mobilization is crucial for bringing troops to the battle's front lines successfully.

no man's land: The territory between enemy trenches. Distances can vary from twenty yards to a mile.

over the top: When a leader in the trenches blows his whistle, soldiers climb out of the trench onto the battlefield.

pillbox: a small, low-lying type of fortress made of reinforced concrete that encloses machine guns and other firearms.

pincer movement: When military forces are sent out in two flanking movements, left and right—like an open claw—before it closes around the enemy, trapping them.

raid: a sudden attack on the enemy. Raids can be conducted by airplanes or by small forces on land.

reconnaissance: getting information about an enemy's position, activities, and resources.

salvo: a simultaneous or one-after-the-other discharge of artillery, bombs, and other munitions.

shrapnel: a hollow projectile that contains bullets or other ammunition and a bursting charge, designed to explode before reaching the target and release a shower of projectiles. Also can be the definition of shell fragments.

shell shock: A type of battle fatigue that causes the human body's nervous system to malfunction.

trenches: a long, narrow depression in the ground that provides shelter and protection from enemy fire or attack.

trench coats: A long waterproof overcoat with a belt and straps on the shoulders and lower sleeves, worn by military officers to protect their clothing in wet trenches.

trench fever: A fever caused by the bites of small insects. Symptoms include fever, weakness, dizziness, headaches, severe back and leg pain, and rashes.

trench foot: Rotting flesh caused by prolonged exposure to cold and wet.

trench mouth: blistering and bleeding gums, mouth, and throat caused by too much bad bacteria.

trip wire: A wire used to set off concealed explosives, such as one stretched across a footpath that would be activated by an enemy soldier's foot.

BIBLIOGRAPHY

Adams, Gregg: *US Marine vs German Soldier: Belleau Wood 1918*. Oxford, UK: Osprey Publishing, 2018

Adams, Simon. *DK Eyewitness Books: World War I*. New York, N.Y.: DK Publishing, Inc., 2007

The Battle of Verdun: A Captivating Guide, 2019

Bonk, David. *Château Thierry & Belleau Wood 1918: America's baptism of fire on the Marne*. Oxford, UK: Osprey Publishing, 2012

Brazear, Margaret. *Journal of William Brazear: An Eye Witness Account of the Battle of Mons*. Margaret Brazear, 2013

Bruce, Anthony. *Complete Illustrated Companion to the First World War*. London: Michael Joseph Publisher, 1990

Captivating History. *The Battle of Verdun: A Captivating Guide…*, (n.p.): *2019*

Eldridge, Jim. *50 Things You Should Know About the First World War*. London: QED Publishing, 2014

Gallishaw, John. *Trenching At Gallipoli*. (n.p.): Otbe Book Publishing, 2016

Giorello, Joe. *Great Battles for Boys: Ancients to Middle Ages*. (n.p.): Rolling Wheelhouse Publishing, 2018

Grant, R.G. *1001 Battles that Changed The Course of History*. New York, N.Y.: Chartwell Books, 2017

Grant, R. G, DK. *World War I: A definitive visual history*. London: DK, 2014

Habeck, Fritz. *Days of Danger*. New York, NY: Collins, 1968

Hanna, Henry. *The Pals at Suvla Bay* (1917). East Sussex, UK: The Naval & Military Press, 2015

Haythornthwaite, Philip. *Gallipoli 1915: Frontal Assault on Turkey*. Oxford, UK: Osprey Publishing, 2013

Lardas, Mark. *World War I Seaplane and Aircraft Carriers*. Oxford, UK: Osprey Publishing, 2016

Lawrence of Arabia: The Life and Legacy of T.E. Lawrence Ann Arbor, MI.: Charles Rivers Editors, 2014

Littel, McDougal. *World War I (Nextext Historical Reader)*. Boston, MA: McDougal Littel Books, 1999

Livesey, Anthony. *Great Battles of World War*. New York, N.Y.: Macmillan General Reference, 1989

Lomas, David. *Mons 1914: The BEF's Tactical Triumph*. Oxford, UK: Osprey Publishing, 2012

McLachlan, Sean. *The Sinai and Palestine Campaign of World War I*, Ann Arbor, MI.: Charles Rivers Editors, 2017

MacLean, Allistair. *Sterling Point Books: Lawrence of Arabia*. New York, NY: Sterling, 2006

Millar, Simon. *Vienna 1683: Christian Europe Repels the Ottomans*. Oxford, UK: Osprey Publishing, 2008

The Ottoman Empire: A Captivating Guide to the Rise and Fall of the Turkish Empire… Captivating History, 2018

The Red Baron: The Life and Legacy of Manfred von Richthofen. Ann Arbor, MI.: Charles Rivers Editors, 2014

Rivers, Charles. *The Red Baron and Eddie Rickenbacker: The Lives and Legacies of World War I's Most Famous Aces*. Ann Arbor, MI.: Charles Rivers Editors, 2017

Stille, Mark. *British Battlecruiser vs German Battlecruiser: 1914–16*.

Oxford, UK: Osprey Publishing, 2013

Stille, Mark. *British Dreadnought vs German Dreadnought: Jutland 1916*. Oxford, UK: Osprey Publishing, 2012

Strachan, Hew. *The First World War*. New York, N.Y.: Pocket Books, 2006

Tuchman, Barbara. *The Guns of August*. New York, N. Y.: Random House Trade Paperbacks, 2014

U.S. Army Command and General Staff College. *A Regiment Like No Other: The 6th Marine Regiment at Belleau Wood*. US: CreateSpace Independent Publishing Format, 2014

Vansant, Wayne. *The Red Baron: The Graphic History of Richthofen's Flying Circus and the Air War in WWI*. Minneapolis:: Zenith Graphic Histories, 2014

Magazines

Military History magazine. (various issues)

Online / Internet

Battle of Jutland narrated by Admiral Jellicoe's grandson: vimeo.com/162655850?ref=fb-share

Battle of Verdun: time.com/4596494/battle-verdun-photos

Documentary on the Battle of Mons: youtube.com/watch?v=oEFoZsuLRoE

First Division Museum's Col. Robert R. McCormick Research Center, The firstdivisionmuseum.nmtvault.com

Great War Period Document Archive, The: www.gwpda.org

National Archives of Britain, The: www.nationalarchives.gov.uk/first-world-war

National WWI Museum and Memorial, The: www.theworldwar.org/explore/online-collections-database

Sea War Museum of Jutland: www.seawarmuseum.dk/en

T.E. Lawrence (Quotes): www.azquotes.com/author/8582-T_E_Lawrence

Turkish warriors' special bow-and-arrow (Video demonstrations of) defense-and-freedom.blogspot.com/2011/04/exotic-ancient-weapons-i-majra.html

US Military Academy map of the Battle of Verdun: www.emersonkent.com/map_archive/battle_of_verdun.htm

US WWI Centennial Commission. WWI aviation timeline www.worldwar1centennial.org/1181-timeline-of-wwi-aviation-history-demo.html

WWI (interactive timeline of): www.abmc.gov/sites/default/files/interactive/interactive_files/WW1/index.html

WWI cavalry, including the Cossacks on the Eastern Front & Wolf packs attacking wounded soldiers: youtube.com/watch?v=XovnkqJaqL8

WWI aircraft and some dogfights (Vintage footage) youtube.com/watch?v=dwrIf_5gEEM

WWI footage from the Battle of Gallipoli: youtube.com/watch?v=DxTKqldkyik

Movies

Battle of Jutland (2016) documentary commemorating the 100-year anniversary of the battle.

Devil Dogs: Hero Marines of WWI

Dog Fight: The Mystery of the Red Baron (2017)

Gallipoli (1999)

Gallipoli (2005) documentary

History's Great Military Blunders and the Lessons They Teach
(2015 video)

Lawrence of Arabia

World War I: American Legacy

Photos

Good Free Photos: www.goodfreephotos.com

Shutterstock: www.shutterstock.com